State-of-the-Art Liquid Crystals Research in Japan

State-of-the-Art Liquid Crystals Research in Japan

Editors

**Shigeyuki Yamada
Kyosuke Isoda
Takahiro Ichikawa
Kosuke Kaneko
Mizuho Kondo
Tsuneaki Sakurai
Atsushi Seki
Mitsuo Hara
Go Watanabe**

Basel • Beijing • Wuhan • Barcelona • Belgrade • Novi Sad • Cluj • Manchester

Editors

Shigeyuki Yamada
Kyoto Institute of Technology
Kyoto, Japan

Kyosuke Isoda
Sagami Chemical Research Institute
Ayase, Japan

Takahiro Ichikawa
Tokyo University of Agriculture and Technology
Tokyo, Japan

Kosuke Kaneko
Ritsumeikan University
Shiga, Japan

Mizuho Kondo
University of Hyogo
Hyogo, Japan

Tsuneaki Sakurai
Kyoto Institute of Technology
Kyoto, Japan

Atsushi Seki
Tokyo University of Science
Tokyo, Japan

Mitsuo Hara
Nagoya University
Aichi, Japan

Go Watanabe
Kitasato University
Kanagawa, Japan

Editorial Office
MDPI
St. Alban-Anlage 66
4052 Basel, Switzerland

This is a reprint of articles from the Special Issue published online in the open access journal *Crystals* (ISSN 2073-4352) (available at: https://www.mdpi.com/journal/crystals/special_issues/liquid_crystals_in_Japan).

For citation purposes, cite each article independently as indicated on the article page online and as indicated below:

Lastname, A.A.; Lastname, B.B. Article Title. *Journal Name* **Year**, *Volume Number*, Page Range.

ISBN 978-3-0365-9548-1 (Hbk)
ISBN 978-3-0365-9549-8 (PDF)
doi.org/10.3390/books978-3-0365-9549-8

© 2023 by the authors. Articles in this book are Open Access and distributed under the Creative Commons Attribution (CC BY) license. The book as a whole is distributed by MDPI under the terms and conditions of the Creative Commons Attribution-NonCommercial-NoDerivs (CC BY-NC-ND) license.

Contents

About the Editors vii

Preface ix

Kumar Siddhant, Ganesan Prabusankar and Osamu Tsutsumi
Luminescent Behavior of Liquid–Crystalline Gold(I) Complexes Bearing a Carbazole Moiety: Effects of Substituent Bulkiness
Reprinted from: *Crystals* **2022**, *12*, 810, doi:10.3390/cryst12060810 1

Atsushi Seki, Kazuki Shimizu and Ken'ichi Aoki
Chiral π-Conjugated Liquid Crystals: Impacts of Ethynyl Linker and Bilateral Symmetry on the Molecular Packing and Functions
Reprinted from: *Crystals* **2022**, *12*, 1278, doi:10.3390/cryst12091278 11

Takahiro Ichikawa, Mei Kuwana and Kaori Suda
Chromonic Ionic Liquid Crystals Forming Nematic and Hexagonal Columnar Phases
Reprinted from: *Crystals* **2022**, *12*, 1548, doi:10.3390/cryst12111548 27

Yuki Arakawa, Yuto Arai, Kyohei Horita, Kenta Komatsu and Hideto Tsuji
Twist–Bend Nematic Phase Behavior of Cyanobiphenyl-Based Dimers with Propane, Ethoxy, and Ethylthio Spacers
Reprinted from: *Crystals* **2022**, *12*, 1734, doi:10.3390/cryst12121734 39

Shigeyuki Yamada, Mitsuki Kataoka, Keigo Yoshida, Masakazu Nagata, Tomohiro Agou, Hiroki Fukumoto and Tsutomu Konno
Development of Hydrogen-Bonded Dimer-Type Photoluminescent Liquid Crystals of Fluorinated Tolanecarboxylic Acid
Reprinted from: *Crystals* **2022**, *13*, 25, doi:10.3390/cryst13010025 51

Mitsuo Hara, Ayaka Masuda, Shusaku Nagano and Takahiro Seki
Photoalignment and Photofixation of Chromonic Mesophase in Ionic Linear Polysiloxanes Using a Dual Irradiation System
Reprinted from: *Crystals* **2023**, *13*, 326, doi:10.3390/cryst13020326 65

Masahiro Ito, Eriko Fukuda, Mitsuhiro Akimoto, Hikaru Hoketsu, Yukitaka Nakazono, Haruki Tohriyama and Kohki Takatoh
Angular Dependence of Guest–Host Liquid Crystal Devices with High Pretilt Angle Using Mixture of Vertical and Horizontal Alignment Materials
Reprinted from: *Crystals* **2023**, *13*, 696, doi:10.3390/cryst13040696 77

Mizuho Kondo, Satoka Yanai, Syouma Shirata, Takeshi Kakibe, Jun-ichi Nishida and Nobuhiro Kawatsuki
Multichromic Behavior of Liquid Crystalline Composite Polymeric Films
Reprinted from: *Crystals* **2023**, *13*, 786, doi:10.3390/cryst13050786 87

Haruka Ohsato, Shigeyuki Yamada, Motohiro Yasui and Tsutomu Konno
Effects of Tetrafluorocyclohexa-1,3-Diene Ring Position on Photoluminescence and Liquid-Crystalline Properties of Tricyclic π-Conjugated Molecules
Reprinted from: *Crystals* **2023**, *13*, 1208, doi:10.3390/cryst13081208 99

Tsuneaki Sakurai, Kenichi Kato and Masaki Shimizu
Side-Chain Labeling Strategy for Forming Self-Sorted Columnar Liquid Crystals from Binary Discotic Systems
Reprinted from: *Crystals* **2023**, *13*, 1473, doi:10.3390/cryst13101473 115

About the Editors

Shigeyuki Yamada

Shigeyuki Yamada, born in Osaka, Japan, in 1980, received his PhD (Engineering) degree from the Kyoto Institute of Technology (2008). After working as a postdoc (2008–2013), he became an Assistant Professor at Ritsumeikan University in 2014. In 2016, he moved to the Kyoto Institute of Technology as an Assistant Professor, and in 2019, he was promoted to Associate Professor. His current research interests include the development of functional materials (e.g., liquid crystals and light-emitting materials) whose functions can be enhanced by introducing fluorine atoms.

Kyosuke Isoda

Kyosuke Isoda, born in Saitama, Japan, in 1981, received his PhD (Engineering) degree from The University of Tokyo (2009). After working as a postdoc (2009–2011) at RIKEN, he worked as an Assistant Professor at the Tokyo University of Science (2011–2014). In 2015, he moved to Kagawa University as a Lecturer, and he was promoted to Associate Professor in 2019. In 2022, he moved to the Sagami Chemical Research Institute, and he was promoted to Group Leader of the Organic Materials Chemistry Group. His current research interests include the development of stimuli-responsive functional materials of liquid materials and liquid crystals.

Takahiro Ichikawa

Takahiro Ichikawa, born in Tokyo, Japan, in 1982, received his PhD (Engineering) degree from The University of Tokyo (2013). After working as an assistant professor at the Tokyo University of Agriculture and Technology (2010–2015), he was promoted to Associate Professor at the same university in 2015. His current research interests include the development of functional gyroid nanostructured materials based on the molecular design of liquid crystals.

Kosuke Kaneko

Kosuke Kaneko, born in Kyoto, Japan, in 1979, received his PhD (Engineering) degree from Ritsumeikan University (2007). After working as a postdoctoral fellowship at the Strasbourg Institute of Material Physics and Chemistry (IPCMS), Centre national de la recherche scientifique (CNRS) (2008–2009), and as a specially appointed Assistant Professor at Kyushu University (2009–2013), he became an Assistant Professor at Ritsumeikan University in 2013. After working on a postdoctoral fellowship at the Fukuoka Institute of Technology (2018), he returned to Ritsumeikan University as an Assistant Professor in 2019. His main activities focus on the design and synthesis of siloxane-based liquid crystalline materials and the investigation of rheological properties by controlling the liquid crystalline organization with an electric field. His recent interests include the development of deformable liquid crystalline elastomers using dual-frequency liquid crystals for creating soft actuator materials.

Mizuho Kondo

Mizuho Kondo studied polymer chemistry at the Tokyo Institute of Technology and completed a Ph.D. under Prof. Tomiki Ikeda in 2009. Thereafter, he worked at the University of Hyogo as an Assistant Professor and was promoted to an Associate Professor in 2020. His research interests include photofunctional organic materials, including polymers and LCs.

Tsuneaki Sakurai

Tsuneaki Sakurai received his PhD in 2012 from the University of Tokyo, Japan. He worked at Osaka University and Kyoto University as a postdoctoral fellow (2012–2014), Assistant Professor (2014–2019), and Junior Associate Professor (2019). After being selected as a member of the Leading Initiative for Excellent Young Researchers program, MEXT, Japan, in 2019, he was appointed as a Junior Associate Professor at the Kyoto Institute of Technology in 2020. His research interests involve the design, synthesis, and evaluation of the physical properties of luminescent organic materials, liquid crystals, conjugated polymers, and various supramolecular systems. He has received several awards, including The Japanese Liquid Crystal Society Young Researcher's Award (2021) and the Award for the Encouragement of Research in Polymer Science (2022).

Atsushi Seki

Atsushi Seki was born in Tokyo, Japan. He graduated from the Department of Applied Chemistry, Faculty of Science and Engineering, Chuo University, in 2007. After he received Master's degree from the University of Tokyo in 2009 under the supervision of Professor Takashi Kato, he worked as an R&D staff member at Toho Chemical Industry Co., Ltd., in Japan from 2009 to 2014. In October, 2014, he joined the research group of Prof. Masahiro Funahashi in the Department of Advanced Materials Science, Faculty of Engineering, Kagawa, as a Ph.D. student. He received his Ph.D. degree from Kagawa University in September, 2017. After he worked as a researcher at Kagawa University for two months, he joined the research group of Professor Nobuyuki Tamaoki in the Research Institute for Electronic Science, Hokkaido University. He moved to the research group directed by Dr. Masafumi Yoshio in the National Institute for Materials Science as a NIMS postdoctoral fellow in February 2018. He became an Assistant Professor at the Tokyo University of Science in 2020. He received the Best Poster Award in the 16th International Conference on Ferroelectric Liquid Crystals (FLC16) in 2016. His current research interests are the development of functional soft matters including ferroelectric π-conjugated liquid crystals, stimuli-responsive supramolecular gels, and photopolymers.

Mitsuo Hara

Mitsuo Hara received his B.S. (2005), M.S. (2007), and Ph.D. (2012) from Nagoya University. From 2007 to 2009, he worked at FUJIFILM Corporation. He was a JSPS Research Fellow in 2010–2012. He started his academic career at Nagoya University as an Assistant Professor. His research interests include self-assembly materials, humidity-responsive polymers, liquid crystalline materials, organic–inorganic hybrid materials, and polymer surfaces. He received The JLCS Award for the Encouragement of Research in Liquid Crystal Science, The JLCS Outstanding Paper Award in 2015 and 2018, and The SPSJ Award for the Encouragement of Research in Polymer Science in 2018.

Go Watanabe

Go Watanabe received his Ph.D. degree from the Department of Pure and Applied Physics, Graduate School of Advanced Science and Engineering, Waseda University, in 2011 under the supervision of Prof. Yuka Tabe. He joined the Department of Physics, School of Science, Kitasato University, in 2012 as an Assistant Professor. Since 2023, he has been a Professor of the Department of Data Science, School of Frontier Engineering, Kitasato University. His research aims to understand the structure and dynamics of liquid crystals, functional organic materials, polymers, and biomolecules with computational science based on molecular simulations.

Preface

Liquid crystals (LCs) are substances that have liquid properties (fluidity) and exhibit the optical anisotropy of crystals, as well as properties that are intermediate between crystals and liquids. LC molecules mainly consist of a rigid π structure and a flexible chain unit, which maintain some directional order (orientation), but do not have a positional order. With thermotropic or lyotropic phase transitions from crystal ⇌ LC ⇌ liquid, the molecular aggregates could be reversibly arranged and expected to appear in molecular dynamic simulations. Altering molecular aggregates is well known to significantly affect their various physical behaviors, including the photophysical, optical, electrical, and chemical aspects. Therefore, there is no doubt that consolidating the results of cutting-edge liquid crystal research into a single Special Issue will greatly contribute to future development within the fields of chemistry, materials, optics, and electronics. This Special Issue, titled "State-of-the-Art Liquid Crystals Research in Japan" is intended to provide an innovative and broad perspective on the LC research taking place in Japan within the fields of chemistry, physics, optics, photonics, photo-alignment techniques, and material and devices, among others.

Shigeyuki Yamada, Kyosuke Isoda, Takahiro Ichikawa, Kosuke Kaneko, Mizuho Kondo, Tsuneaki Sakurai, Atsushi Seki, Mitsuo Hara, and Go Watanabe
Editors

Article

Luminescent Behavior of Liquid–Crystalline Gold(I) Complexes Bearing a Carbazole Moiety: Effects of Substituent Bulkiness

Kumar Siddhant [1], Ganesan Prabusankar [2] and Osamu Tsutsumi [1,*]

[1] Department of Applied Chemistry, College of Life Sciences, Ritsumeikan University, Kusatsu 525-8577, Japan; gr0451pe@ed.ritsumei.ac.jp

[2] Department of Chemistry, Indian Institute of Technology Hyderabad, Kandi 502285, India; prabu@chy.iith.ac.in

* Correspondence: tsutsumi@sk.ritsumei.ac.jp

Abstract: Organometallic materials that exhibit white luminescence in condensed phases are of considerable interest for lighting and display applications. Herein, new carbazole-based Au(I) complexes containing an isocyanide group and a long pentyl chain were synthesized. The complex with an unsubstituted carbazole moiety exhibited a white emission at room temperature as well as nematic liquid crystalline behavior. Color tunability from white to blue was achieved when bulkier substituents were introduced at the 3 and 6 positions of the carbazole moiety. Furthermore, all complexes possessed long phosphorescence lifetimes in the crystal state. The proposed design framework provides new opportunities for practical applications using luminescent organometallic molecules.

Keywords: white emission; liquid crystal; aggregated structure; gold(I) complex; carbazole moiety

1. Introduction

Luminescent molecules are of importance in both materials and life sciences [1–3]. In particular, white-light-emitting organometallic materials have attracted increasing attention owing to their potential applications in lighting devices and display media. However, the generation of white light commonly requires the simultaneous emission of the three primary RGB colors (red, green, and blue) or at least two complementary colors [4]. Moreover, achieving white room temperature phosphorescence (RTP) from a single molecule is highly challenging because the different radiation pathways involved in the emission process can affect each other [5].

Although most luminescent organometallic molecules exhibit efficient photoluminescence in dilute solutions, aggregation in condensed phases (e.g., crystals and solid films) typically results in partial or complete quenching of their luminescence. This phenomenon, known as aggregation-caused quenching (ACQ), is common in organometallic molecules with π-electron systems and prevents their practical use. Owing to the essential contribution of molecular aggregation to luminescence properties, aggregated structures play an important role in luminescence behavior [5–10]. Liquid crystals (LCs) are a unique class of soft materials that flow like liquids and possess a long-range orientational order similar to that of crystals [7,11–14]. LCs have the potential to control the aggregated structure of luminescent materials. Moreover, the material properties and aggregated structures of LC molecules can be influenced by various external stimuli. The molecular skeleton of an LC molecule generally consists of a rod-like rigid core and flexible chains, such as alkoxy and alkyl chains. Various aggregation-induced emission (AIE)-active LC systems have been developed [15,16].

Au(I) complexes show very interesting behavior because of their metallophilic d^{10}–d^{10} interactions. The emission observed in Au(I) complexes results from an interatomic Au–Au interaction; thus, Au(I) complexes exhibit the luminescence mainly from their aggregates

Citation: Siddhant, K.; Prabusankar, G.; Tsutsumi, O. Luminescent Behavior of Liquid–Crystalline Gold(I) Complexes Bearing a Carbazole Moiety: Effects of Substituent Bulkiness. *Crystals* **2022**, *12*, 810. https://doi.org/10.3390/cryst12060810

Academic Editor: Paul R. Raithby

Received: 12 May 2022
Accepted: 7 June 2022
Published: 8 June 2022

Publisher's Note: MDPI stays neutral with regard to jurisdictional claims in published maps and institutional affiliations.

Copyright: © 2022 by the authors. Licensee MDPI, Basel, Switzerland. This article is an open access article distributed under the terms and conditions of the Creative Commons Attribution (CC BY) license (https://creativecommons.org/licenses/by/4.0/).

and have been known as the AIE-active material. The Au–Au interaction as well as the luminescence behavior can be tuned via not only modification of the chemical structure around Au atoms but also by controlling the aggregated structures [5]. For instance, previously synthesized Au(I) complexes with isocyanides and phenylacetylene ligands exhibited interesting AIE properties [16]. Furthermore, our group reported a biphenyl ring system with an LC phase, where luminescence was observed in both the crystal and LC phases [14].

Carbazole-based luminescent materials are valuable candidates for photoelectronic devices. Carbazole units are suitable building blocks for functional materials in organometallic light-emitting diodes because of their high triplet energy and good thermal stability [17]. Herein, we developed a carbazole-based Au(I) complex with a flexible pentyl isocyanide group. This complex shows a white emission at room temperature (rt) along with a nematic (N) LC phase. Color tunability from white (**Cbz-H**) to blue (**Cbz-*t*-Bu**) was also realized by changing the substituents at the 3 and 6 positions of the carbazole moiety.

2. Materials and Methods

2.1. Materials

Compound **1-R** was prepared by following the reported procedure [18]. Complexes **Cbz-H**, **Cbz-Br**, and **Cbz-*t*-Bu** were prepared via a one-step synthetic route using compounds **1-R**. The reagents and solvents used for the synthesis were obtained from commercial sources and used without further purification. ^{1}H and ^{13}C NMR spectra were recorded using a JEOL ECS-400 spectrometer (JEOL, Tokyo, Japan) (400 MHz for ^{1}H and 100 MHz for ^{13}C) in CDCl$_3$ (Figures S1–S3 for ^{1}H NMR, Figures S4–S6 for ^{13}C NMR). Chemical shifts are reported in parts per million (ppm), using the residual ^{1}H or ^{13}C in the NMR solvent as an internal reference. IR spectra were obtained using the KBr disk method with a FT/IR-4100 spectrometer (JASCO, Tokyo, Japan), and all spectra are reported in wavenumbers (cm^{-1}). Elemental analysis (C, H, and N) was performed using a Micro Corder JM10 analyzer (J-Science, Tokyo, Japan). Electrospray ionization mass spectrometry (ESI-MS) was carried out using a Bruker micrOTOF II instrument (JEOL, Tokyo, Japan).

2.1.1. Synthesis of Cbz-H

Compound **1-H** (0.30 g, 1.5 mmol) and (tht) AuCl (tht: tetrahydrothiophene) (0.47 g, 1.5 mmol) were dissolved in 10 mL of CH$_2$Cl$_2$. Then, a methanol (10 mL) solution of CH$_3$COONa (0.61 g, 7.2 mmol) was added dropwise, and the resulting mixture was stirred for 2 h at rt. Subsequently, 1-pentyl isocyanide (0.071 g, 0.73 mmol) in CH$_2$Cl$_2$ (7 mL) was added and the mixture was stirred for a further 1 h at rt. The reaction mixture was then filtered through Celite and evaporated under vacuum. The crude product was purified by silica gel column chromatography using CH$_2$Cl$_2$ as the mobile phase and then recrystallized in a CH$_2$Cl$_2$/*n*-hexane mixed solvent system (1:2, *v*/*v*) to obtain 0.28 g, 0.57 mmol of the title complex as white crystal in 78% yield, mp 156 °C. ^{1}H NMR (400 MHz, CDCl$_3$, δ): 8.06–8.04 (d, *J* = 7.5 Hz, 2H, Ar*H*), 7.56–7.54 (dd, *J* = 7.5 Hz, 2H, Ar*H*), 7.44 (d, *J* = 7.5 Hz, 2H, Ar*H*), 7.25–7.20 (m, *J* = 7.5 Hz, 2H, Ar*H*), 5.15–5.12 (s, 2H, ArNC*H*), 3.52–3.49 (t, 2H, NC*H*), 1.73 (s, *J* = 7.1 Hz, 2H, C*H*$_2$), 1.36–1.2 (m, *J* = 7.1 Hz, 2H, C*H*$_2$), 0.95–0.87 (t, *J* = 8 Hz 3H, C*H*$_3$). ^{13}C NMR (400 MHz, CDCl$_3$, δ): 140.04 (9-C in carbazole), 125.78 (2,7-C in carbazole), 123.09 (Au–C≡N), 120.23 (3,6-C in carbazole), 119.10 (4-C in carbazole), 114.84 (4-C in carbazole), 109.29 (1,8-C in carbazole), 77.13 (Au–C≡C), 43.83 (N–CH$_2$), 33.44 (N–CH$_2$ in carbazole), 28.22 (NCH$_2$CH$_2$), 27.56 (NCH$_2$CH$_2$CH$_2$), 21.71 (CH$_2$CH$_3$), 13.79 (CH$_2$CH$_3$). FTIR (KBr, cm^{-1}): 3049, 2953, 2248, 1460. ESI-MS *m*/*z*: [M + H]$^+$ calcd for C$_{21}$H$_{21}$AuN$_2$, 499.14; found, 499.11. Anal calcd for C$_{21}$H$_{21}$AuN$_2$: C, 50.61; H, 4.25; Au, 39.52; N, 5.62; found: C, 50.43; H, 4.12; N, 5.51.

2.1.2. Synthesis of Cbz-Br

The same procedure was followed using **1-Br** for the synthesis of **Cbz-Br** to obtain the title complex as a white solid in 84% yield, mp 168 °C. ^{1}H NMR (400 MHz, CDCl$_3$, δ): 8.11–8.10 (d, J = 7.5 Hz, 2H, ArH), 7.56–7.53 (dd, J = 7.5 Hz, 2H, ArH), 7.44–7.42 (d, J = 7.5 Hz, 2H, ArH), 5.09 (s, 2H, ArNCH), 3.56–3.52 (s, 2H, NCH), 1.76 (s, J = 7.1 Hz, 2H, CH_2), 1.39–1.32 (m, J = 7.1 Hz, 2H, CH_2), 0.92–0.89 (t, J = 8 Hz 3H, CH_3). ^{13}C NMR (400 MHz, CDCl$_3$, δ): 139.05 (9-C in carbazole), 129.13 (2,7–C in carbazole), 123.84 (Au–C≡N), 123.17 (3,6-C–Br in carbazole), 115.72 (4-C in carbazole), 112.34 (4-C in carbazole), 111.13 (1,8-C in carbazole), 77.11 (Au–C≡C), 40.01 (N–CH$_2$), 33.79 (N–CH$_2$ in carbazole), 33.41 (NCH$_2$CH$_2$), 27.77 (NCH$_2$CH$_2$CH$_2$), 21.75 (CH$_2$CH$_3$), 13.80 (CH$_2$CH$_3$). FTIR (KBr, cm^{-1}): 3073, 2957, 2245, 1472, 645. ESI-MS m/z: [M]$^+$ calcd for C$_{21}$H$_{19}$AuBr$_2$N$_2$, 653.96; found, 653.46. Anal calcd for C$_{21}$H$_{19}$AuBr$_2$N$_2$: C, 38.44; H, 2.92; Au, 30.02; N, 4.27; Br, 24.35 found: C, 38.27; H, 2.79; N, 4.21.

2.1.3. Synthesis of Cbz-t-Bu

The same procedure was followed using **1-t-Bu** for the synthesis of **Cbz-t-Bu** to obtain the title complex as white solid in 81% yield, mp 190 °C. ^{1}H NMR (400 MHz, CDCl$_3$, δ): 8.04 (d, J = 7.5 Hz, 2H, ArH), 7.50–7.47 (t, J = 7.5 Hz, 2H, ArH), 7.45–7.42 (d, J = 7.5 Hz, 2H, ArH), 5.09 (s, 2H, ArNCH), 3.43–3.40 (t, 2H, NCH), 1.53 (s, J = 7.1 Hz, 2H, CH_2), 1.42 (s, 18H, CH_3), 1.29–1.28 (m, J = 7.1 Hz, 4H, CH_2), 0.87 (t, J = 8 Hz 3H, CH_3). ^{13}C NMR (400 MHz, CDCl$_3$, δ): 141.72 (C(CH$_3$)$_3$), 138.60 (9-C in carbazole), 123.42 (Au–C≡N), 122.98 (2,7-C in carbazole), 116.17 (4-C in carbazole), 114.45 (4a-C in carbazole), 108.59 (1,8-C in carbazole), 77.12 (Au–C≡C), 43.86 (N–CH$_2$), 34.74 (N–CH$_2$ in carbazole), 33.41 (NCH$_2$CH$_2$), 32.16 (C(CH$_3$)$_3$), 27.61 (NCH$_2$CH$_2$CH$_2$), 21.729 (CH$_2$CH$_3$), 13.79 (CH$_2$CH$_3$). FTIR (KBr, cm^{-1}): 3044, 2953, 2245, 1477. ESI-MS m/z: [M + H]$^+$ calcd for C$_{29}$H$_{37}$AuN$_2$, 611.27; found, 611.26. Anal calcd for C$_{29}$H$_{37}$AuN$_2$: C, 57.05; H, 6.11; Au, 32.26; N, 4.59; found: C, 56.94; H, 6.05; N, 4.37.

2.2. X-ray Crystallography

A single crystal was prepared via slow evaporation in a mixed solvent system (CH$_2$Cl$_2$/ n-hexane). The obtained crystal was mounted on a glass fiber. The omega scanning technique was applied to collect the reflection data using a Bruker D8 goniometer (Bruker, Millerica, MA, USA) with monochromatized MoKα radiation (λ = 0.71075 Å). To estimate the actual crystal structure of the synthesized materials, measurements were performed at ambient temperature (296 K). The initial structure of the unit cell was determined via a direct method using APEX2. The structural model was refined by the full-matrix least-squares method using SHELXL-2014/6. All calculations were performed using the SHELXL software [19,20]. The crystallographic data for the synthesized compounds are summarized in the Supplementary Materials, and the indexed data were deposited in the Cambridge Crystallographic Data Centre (CCDC) database (CCDC 2160379 for **Cbz-H**). These data can be obtained free of charge via http://www.ccdc.cam.ac.uk/conts/retrieving.html accessed on 21 March 2022 (or from the CCDC, 12 Union Road, Cambridge CB2 1EZ, UK; Fax: +44-1223-336033; E-mail: deposit@ccdc.cam.ac.uk).

2.3. Phase Transition Behavior

LC behavior as well as the melting points of the complexes were observed via polarized optical microscopy (POM) using an Olympus BX51 (Tokyo, Japan) microscope equipped with a hot stage (Instec HCS302 hot stage with an mK1000 controller). To assess the thermochemical stability, thermogravimetric–differential thermal analysis (TG/DTA) was carried out using a DTG-60AH instrument (Shimadzu, Kyoto, Japan) at a heating rate of 5.0 °C min^{-1} in air. The thermodynamic parameters were determined via differential scanning calorimetry (DSC; SII X-DSC7000, Tokyo, Japan) at a scanning rate of 3.0 °C min^{-1}.

2.4. Photophysical Properties

UV–Vis absorption spectra were recorded on a JASCO V-550 (JASCO, Tokyo, Japan) absorption spectrometer, and steady-state photoluminescence spectra were recorded on a Hitachi F-7000 fluorescence spectrometer (Hitachi, Tokyo, Japan) with an R928 photomultiplier (Hamamatsu Photonics, Hamamatsu, Japan) as the detector. Photoluminescence lifetimes were measured at various excitation wavelengths using a Quantaurus-Tau photoluminescence lifetime measurement system (C1136-21, Hamamatsu Photonics, Hamamatsu, Japan).

3. Results and Discussion

3.1. Synthesis and Characterization of Complex Cbz-R

Complex **Cbz-R** were synthesized according to the synthetic route shown in Figure 1. After purification, all complexes were characterized via ^{1}H NMR spectroscopy, mass spectrometry, and elemental analysis. All the analytical data (presented in the Materials and Methods section) confirmed that the desired products were obtained.

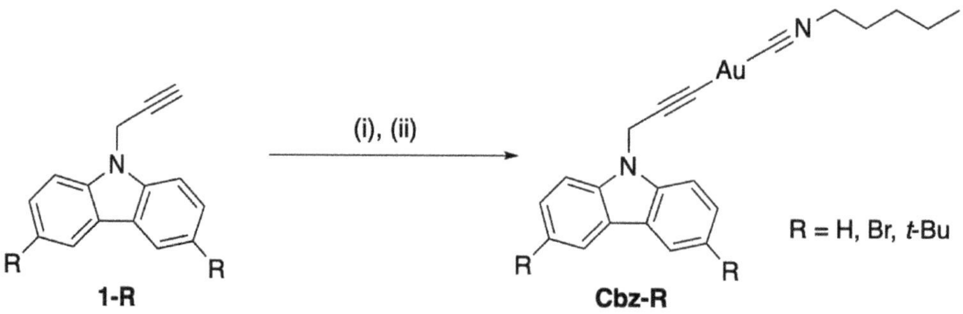

Figure 1. Molecular structure and synthetic route for complex **Cbz-R**: (i) (tht) AuCl/CH$_2$Cl$_2$, CH$_3$COONa/MeOH; (ii) 1-pentyl isocyanide.

Complex **Cbz-H** furnished single crystals suitable for X-ray crystallography via a slow evaporation technique using the mixed solvent system; however, suitable single crystals were not obtained for other complexes. To determine the molecular structure of **Cbz-H**, single-crystal X-ray diffraction (XRD) was performed at room temperature. The obtained crystal structure is shown in Figure 2. The corresponding crystallographic data and key structural parameters are summarized in Tables S1 and S2, respectively. According to the crystal structure, **Cbz-H** belongs to the triclinic space group $P\bar{1}$. The C1–Au–C2 bond angle is ~180° [21], and molecular packing gives a Au–Au bond distance of 6.4 Å, which suggests the absence of aurophilic interactions between the molecules. We have previously reported the structure of a similar Au(I) complex that has less-hindered 4-ethynylbiphenyl and alkyl isocyanide ligands [14]. This complex also crystallized in triclinic space group $P\bar{1}$, and the C1–Au–C2 bond angles were the same as in **Cbz-H**, i.e., ~180°; however, the intermolecular Au–Au distance was 3.46 Å, which confirms the presence of intermolecular aurophilic interactions in the crystal. The longer intermolecular distance between Au atoms in **Cbz-H** is due to the bulkier carbazolylmethyl unit. Additionally, the face of the carbazole moiety is orthogonal to the C1–Au–C2 axis. We consider that this orthogonal structure of the bulky carbazole moiety resulted in the longer intermolecular Au–Au distance in the **Cbz-H** complex.

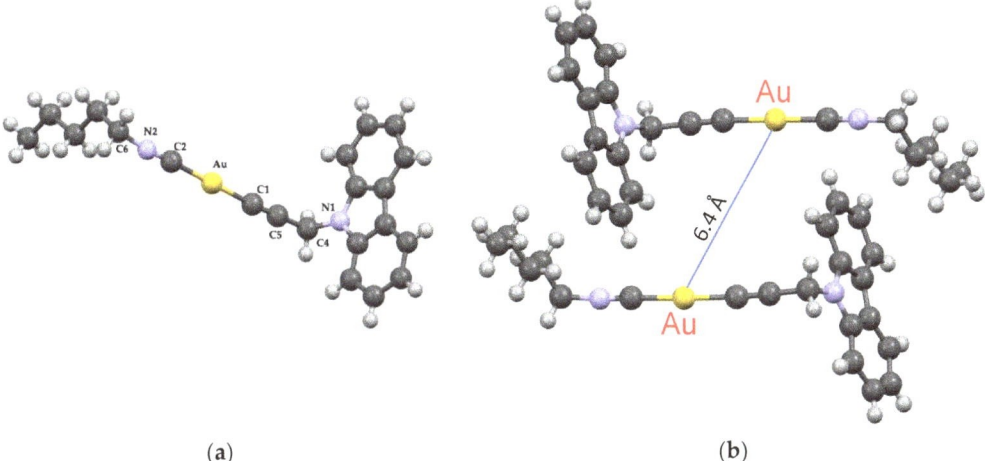

Figure 2. Molecular structure (**a**) and molecular packing structure (**b**) of complex **Cbz-H** determined via single-crystal X-ray diffraction at rt (gray, C; blue, N; yellow, Au; white, H).

3.2. Thermal Behavior of Complex Cbz-R

The thermal stability of complex **Cbz-R** was analyzed using TG/DTA measurements (Figure S7). The thermal decomposition temperature (T_{dec}) is defined as the temperature at which 5% weight loss occurs [22,23]. The TG/DTA thermograms show that all of the synthesized complexes are thermally stable up to 170–220 °C. The 12–15% weight loss observed near T_{dec} can be attributed to the loss of the isocyanide moieties. The second weight loss may be due to the removal of the propargyl moiety. The amount of residual ash obtained at 600 °C corresponds to the percentage of gold(I) in the complex.

The thermodynamic behavior was evaluated using DSC and POM techniques. The DSC thermogram of **Cbz-H** suggests LC behavior during heating (Figure S8). To further investigate this behavior, POM images were collected to examine the optical texture of **Cbz-H** (Figure 3). A schlieren texture was obtained at 93 °C, which confirmed that the observed phase for **Cbz-H** is the N phase [14,16]. In the DSC thermogram, a small endothermic peak was observed at 73 °C during the heating process, and an isotropic phase was achieved at 118 °C. The N phase was observed between these temperatures. However, in the cooling process of the DSC thermograms, no exothermic peak was observed. Through POM observation, we confirmed that the phase transition from the isotropic liquid to the amorphous solid occurred at ~80 °C, and that once **Cbz-H** was melted, the complex did not exhibit crystalline and LC phases.

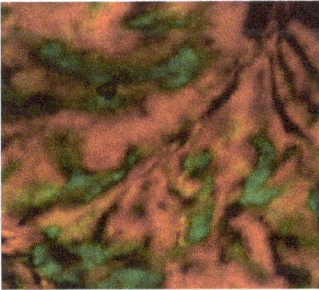

Figure 3. Optical texture of **Cbz-H** observed via polarized optical microscopy at 93 °C during the first heating process (scanning rate: 3.0 °C min^{-1}).

In contrast, no LC phases were observed for complexes **Cbz-Br** and **Cbz-*t*-Bu** at any temperature. In **Cbz-Br** and **Cbz-*t*-Bu**, the substituents attached at the 3 and 6 positions of the carbazole moiety are bulkier than H atoms (i.e., Br atoms or *t*-butyl groups). As the steric hindrance of these bulky substituents disturbs the LC alignment, only **Cbz-H** shows an LC phase [24–26].

3.3. Solution and Solid-State Photoluminescence Properties of Complex Cbz-R

The UV–vis absorption spectra of compounds **Cbz-H**, **Cbz-Br**, and **Cbz-*t*-Bu** were recorded in dilute CH_2Cl_2 solution. **Cbz-H** (2.1×10^{-5} mol L^{-1}), **Cbz-Br** (4.3×10^{-5} mol L^{-1}), and **Cbz-*t*-Bu** (3.3×10^{-5} mol L^{-1}) exhibited absorption bands at 293, 310, and 297 nm, respectively (Figure 4a), which may be attributable to a metal-to-ligand charge transfer (MLCT) transition or a ligand-based π–π* transition [5,27,28]. All complexes also showed smaller absorption peaks with a vibronic structure at 340–360 nm. This small absorption band is known as the n–π* transition of the carbazole unit. Additionally, it has also been reported that the absorption band at this wavelength originated from the transition from the alkynyl-to-isocyanide ligand-to-ligand charge transfer [29]. Thus, we can consider that both transitions are overlapped in this absorption band.

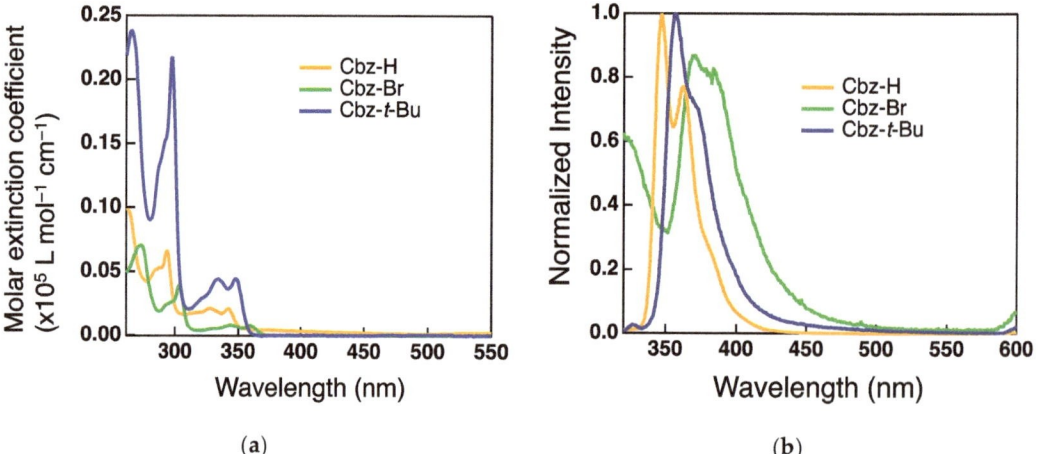

Figure 4. (a) UV–Vis absorption spectra of complex **Cbz-R** in CH_2Cl_2 solution: [**Cbz-H**], 2.1×10^{-5} mol L^{-1}; [**Cbz-Br**], 4.3×10^{-5} mol L^{-1}; [**Cbz-*t*-Bu**], 3.3×10^{-5} mol L^{-1}. (b) Normalized emission spectra of complex **Cbz-R** in CH_2Cl_2 solution: [**Cbz-H**], 2.1×10^{-5} mol L^{-1}, λ_{ex} = 293 nm; [**Cbz-Br**], 4.3×10^{-5} mol L^{-1}, λ_{ex} = 300 nm; [**Cbz-*t*-Bu**], 3.3×10^{-5} mol L^{-1}, λ_{ex} = 297 nm.

All complexes showed monomeric-type emissions with a vibronic structure at ~350 nm in dilute solutions (Figure 4b). Similar spectral shapes were observed at this wavelength in all complexes; however, in **Cbz-*t*-Bu** and **Cbz-Br**, the vibronic bands were slightly broadened. The emission peaks of **Cbz-H** and **Cbz-*t*-Bu** were observed at 375 and 380 nm, respectively, whereas that of **Cbz-Br** was slightly red-shifted to approximately 400 nm. Although the luminescence behaviors of all complexes were comparable in the dilute solution, significant differences were observed in the solid state.

The emission spectra of the complexes were obtained in the crystal state (Figure 5a). Each complex exhibited a peak in the region of 370–400 nm, which is similar to the emission peak in the solution state and can be considered a monomeric-type fluorescence emission. Furthermore, **Cbz-Br** and **Cbz-*t*-Bu** exhibited additional emission peaks at approximately 430 and 450 nm. Thus, the emission of these complexes in the crystal state is more structured than that in the crystal state and extends into the blue region. Furthermore, **Cbz-Br** exhibits a brighter blue emission than **Cbz-*t*-Bu** owing to the heavy-atom effect [30]. In contrast, the emission of **Cbz-H** covers the entire spectral range, including a prominent peak at approximately 550 nm due to crystallization-induced phosphorescence [31–33]. Consequently, **Cbz-H** shows a white emission. As shown by the CIE plot (Figure 5b), **Cbz-H**, **Cbz-Br**, and **Cbz-*t*-Bu** show white, bluish-white, and bluish emissions, respectively. Thus, color tunability from white to blue was successfully achieved by changing the substituents at the 3 and 6 positions of the carbazole moiety. The color tunability of these complexes can be attributed to the introduction of bulky substituents, which inhibits free rotation of the luminophore.

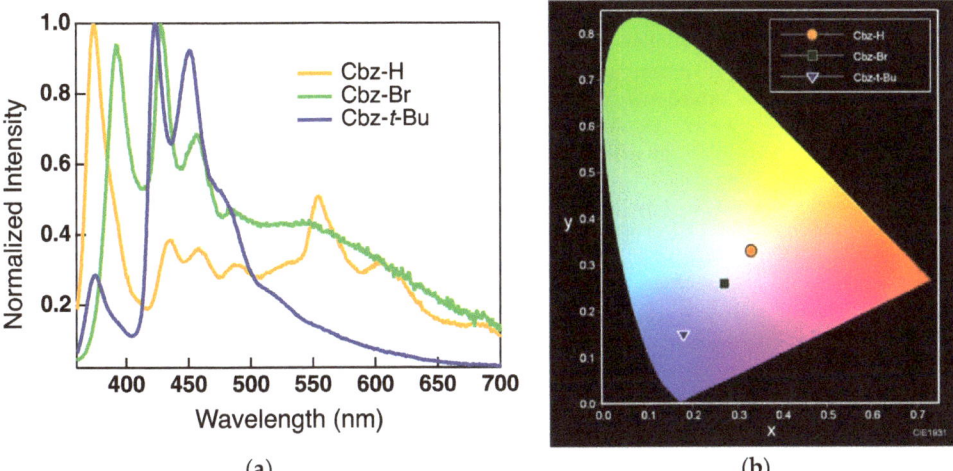

Figure 5. (a) Solid-state emission spectra (λ_{ex} = 300 nm) of complex **Cbz-R** (orange, **Cbz-H**; green, **Cbz-Br**; blue, **Cbz-*t*-Bu**). (b) CIE plot for the luminescence of complex **Cbz-R** in the crystal state.

To further clarify the photophysical properties of these complexes, the lifetimes and luminescence quantum yields were measured in the crystal state at room temperature (Figures S13–S15, Table S3). The decay profiles were well-fitted using a biexponential function. In all complexes, it was confirmed that the emission band at <400 nm was fluorescence due to the S_1–S_0 transition. In **Cbz-H**, the emission at 450 nm decayed on a nanosecond timescale, indicating that the emission bands in this wavelength region can be also assigned to fluorescence emitted from a higher excited state (S_n). In contrast, the emission at 550 nm contains a component with a lifetime of the microsecond timescale, suggesting that a part of this band can be attributed to phosphorescence. Interestingly, in **Cbz-Br**, emissions at 450 nm and 550 nm decayed on a microsecond timescale, and both emission bands can be attributed to phosphorescence due to the heavy-atom effects of the Br atoms. In **Cbz-*t*-Bu**, contrary to **Cbz-H**, phosphorescence was observed at 450 nm; however, fluorescence mainly contributed to the emission at 550 nm. In the **Cbz-*t*-Bu** crystal, we consider that phosphorescence from monomers of the complex was observed at 450 nm, and that fluorescence from the aggregates, i.e., excimer-type fluorescence, was observed at ~550 nm.

As shown in Table S3, the complexes had luminescence quantum yields (Φ) of 0.34–2.3% in the crystal state, even at room temperature in air. The excitation spectra of the crystals (Figure S9) revealed an excitation band at shorter wavelengths (320–340 nm), which is assigned to the S_0–S_n transition. The excitation spectra of all complexes showed an additional band at longer wavelengths (380–430 nm). We previously reported that similar Au(I) complexes with ethynyl and isocyanide ligands showed an efficient direct S_0–T_n transition owing to aggregation; that is, the S_0–T_n transition is enhanced by not only the heavy-atom effect but also the aggregation of Au(I) complexes. Thus, we conclude that the longer-wavelength band in the excitation spectra can be assigned to a spin-forbidden direct S_0–T_n transition. Because the spin-forbidden transition occurs efficiently, the complexes show intense RTP in crystals under ambient conditions [14,34].

The photoluminescence of **Cbz-H** in the LC phase was also examined to determine the effects of the phase structure on luminescence behavior (Figure S10). The emission intensity levels were recorded at various temperatures. The emission intensity gradually decreased upon heating from rt to the LC phase, which indicates that a nonradiative decay was promoted by thermal motion of the Au(I) complex [14,16,22]. The emission intensity was recovered in the cooling process; however, it still remained lower than that of the original crystal before heating. This behavior is attributed to the crystal size dependence of the aggregation-enhanced spin-forbidden transition [14,34].

To obtain further insight into the luminescence properties, photophysical studies were performed in a mixed water–tetrahydrofuran (THF) system with various volume fractions of water. As all compounds showed good solubility in THF and very poor solubility in water, increasing the volume fraction of water induced aggregation. As shown in Figure S11, when the water concentration increased, the luminescence intensity decreased because of molecular aggregation. These results demonstrate that complexes of **Cbz-R** exhibit ACQ behavior [7]. In the case of **Cbz-H**, the same spectral shape was observed in the mixed solvent with a high water fraction as that in pure THF; namely, the spectral shape of aggregates in the mixed solvent was different from those in the crystal and LC, as shown in Figure S10. The results suggest that **Cbz-H** formed aggregates with different aggregate structures from the crystal, and that the luminescence behavior of this complex is sensitive to its aggregated structure, although no change in the spectral shape was observed via crystal-to-LC phase transition.

4. Conclusions

Color tunability was observed for carbazole-based Au(I) complexes, with a color change from white to blue occurring as the steric bulkiness of the carbazole moiety increased. Furthermore, all complexes showed phosphorescence in the crystal state. Notably, complex **Cbz-H** showed an N phase along with a white emission at room temperature. However, no LC phases were observed for **Cbz-Br** and **Cbz-t-Bu**. These findings suggest a new approach for designing luminescent materials.

Supplementary Materials: The following are available online at https://www.mdpi.com/article/10.3390/cryst12060810/s1: Figure S1: ^{1}H NMR spectra of complex **Cbz-H**. Figure S2: ^{1}H NMR spectra of complex **Cbz-Br**. Figure S3: ^{1}H NMR spectra of complex **Cbz-t-Bu**. Figure S4: ^{13}C NMR spectra of complex **Cbz-H**. Figure S5: ^{13}C NMR spectra of complex **Cbz-Br**. Figure S6: ^{13}C NMR spectra of complex **Cbz-t-Bu**. Table S1: Crystallographic data of complex **Cbz-H**. Table S2: Optimized bond parameters for **Cbz-H**. Figure S7: TG/DTA thermograms of complex **Cbz-R**. Figure S8: DSC thermogram of complex **Cbz-H**. Table S3: Photoluminescence decay parameters of complex **Cbz-R**. Figure S9: Excitation spectra of complex **Cbz-R** in solid state. Figure S10: Solid-state emission spectra of complex **Cbz-H** after first heating and cooling. Figure S11: Emission spectra of complex **Cbz-R** in THF–water mixture. Figure S12: Emission spectra of complex **Cbz-R** in CH_2Cl_2 before and after degassing. Figures S13–S15: Photoluminescence decay profile of complex **Cbz-R**.

Author Contributions: Conceptualization, O.T. and G.P.; methodology, K.S.; formal analysis, K.S.; investigation, K.S.; writing—original draft preparation, K.S.; writing—review and editing, O.T. and G.P.; supervision, O.T. and G.P.; project administration, O.T.; funding acquisition, O.T. All authors have read and agreed to the published version of the manuscript.

Funding: This research was supported by the Japan-India Science Cooperative Program between JSPS and DST (JPJSBP120217715), JICA, and the Cooperative Research Program of the Network Joint Research Centre for Materials and Devices.

Institutional Review Board Statement: Not applicable.

Informed Consent Statement: Not applicable.

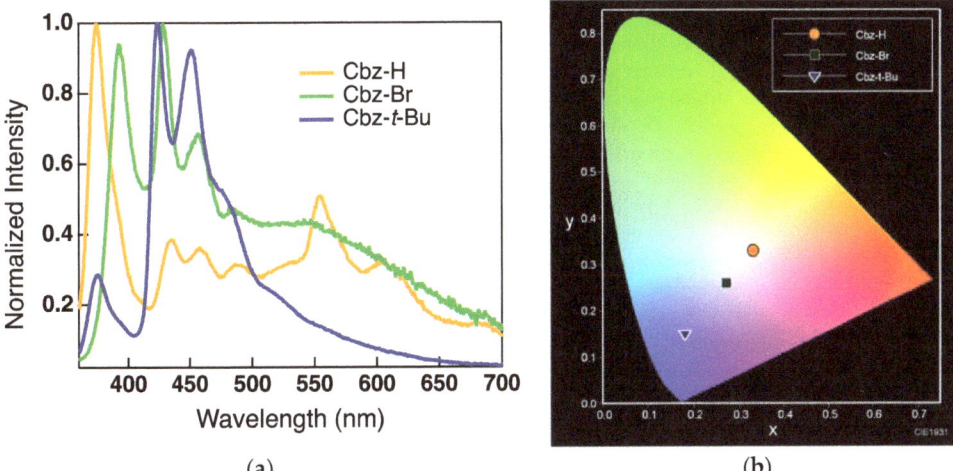

Data Availability Statement: The data presented in this study are available in the article and Supplementary Materials.

Conflicts of Interest: The authors declare no conflict of interest.

References

1. Ghosh, B.; Shirahata, N. Colloidal silicon quantum dots: Synthesis and luminescence tuning from the near-UV to the near-IR range. *Sci. Technol. Adv. Mater.* **2014**, *15*, 014207. [CrossRef] [PubMed]
2. Dubey, V.; Som, S.; Kumar, V. (Eds.) *Luminescent Materials in Display and Biomedical Applications*, 1st ed.; CRC Press: Boca Raton, FL, USA, 2020.
3. Shizu, K.; Lee, J.; Tanaka, H.; Nomura, H.; Yasuda, T.; Kaji, H.; Adachi, C. Highly efficient electroluminescence from purely organic donor–acceptor systems. *Pure Appl. Chem.* **2015**, *87*, 627–638. [CrossRef]
4. Li, D.; Hu, W.; Wang, J.; Zhang, Q.; Cao, X.-M.; Ma, X.; Tian, H. White-light emission from a single organic compound with unique self-folded conformation and multistimuli responsiveness. *Chem. Sci.* **2018**, *9*, 5709–5715. [CrossRef] [PubMed]
5. Sathyanarayana, A.; Siddhant, K.; Yamane, M.; Hisano, K.; Prabusankar, G.; Tsutsumi, O. Tuning the Au–Au interactions by varying the degree of polymerization in linear polymeric Au(I) N-heterocyclic carbene complexes. *J. Mater. Chem. C* **2022**, *10*, 6050–6060. [CrossRef]
6. Birks, J.B. *Photophysics of Aromatic Molecules*; Wiley-Interscience: London, UK, 1970.
7. Sami, H.; Younis, O.; Maruoka, Y.; Yamaguchi, K.; Siddhant, K.; Hisano, K.; Tsutsumi, O. Negative thermal quenching of photoluminescence from liquid-crystalline molecules in condensed phases. *Crystals* **2021**, *11*, 1555. [CrossRef]
8. Ronda, C.R. Emission and excitation mechanisms of phosphors. In *Luminescence: From Theory to Applications*; Ronda, C.R., Ed.; Wiley-VCH: Weinheim, Germany, 2008; pp. 1–34.
9. Xu, J.; Chua, M.H.; Tang, B.Z. (Eds.) *Aggregation-Induced Emission (AIE): A Practical Guide (Materials Today)*, 1st ed.; Elsevier: Amsterdam, The Netherlands, 2022.
10. Tang, Y.; Tang, B.Z. (Eds.) *Handbook of Aggregation-Induced Emission, Vol. 1: Tutorial Lectures and Mechanism Studies*; Wiley: Chichester, UK, 2022.
11. Khoo, I.-C. *Liquid Crystals*, 2nd ed.; Wiley Interscience: Hoboken, NJ, USA, 2007.
12. Schadt, M. Liquid crystal materials and liquid crystal displays. *Annu. Rev. Mater. Sci.* **1997**, *27*, 305–379. [CrossRef]
13. Demus, D.; Goodby, J.; Gray, G.W.; Spiess, H.-W.; Vill, V. (Eds.) *Handbook of Liquid Crystals*; Wiley-VCH: Weinheim, Germany, 1998.
14. Furoida, A.; Daitani, M.; Hisano, K.; Tsutsumi, O. Aggregation-enhanced room-temperature phosphorescence from Au(I) complexes bearing mesogenic biphenylethynyl ligands. *Molecules* **2021**, *26*, 7255. [CrossRef]
15. Yamada, S.; Rokusha, Y.; Kawano, R.; Fujisawa, K.; Tsutsumi, O. Mesogenic gold complexes showing aggregation-induced enhancement of phosphorescence in both crystalline and liquid-crystalline phases. *Faraday Discuss.* **2017**, *196*, 269–283. [CrossRef]
16. Fujisawa, K.; Kawakami, N.; Onishi, Y.; Izumi, Y.; Tamai, S.; Sugimoto, N.; Tsutsumi, O. Photoluminescent properties of liquid crystalline gold(I) isocyanide complexes with a rod-like molecular structure. *J. Mater. Chem. C* **2013**, *1*, 5359–5366. [CrossRef]
17. Chen, Z.; Liu, G.; Pu, S.; Liu, S.H. Carbazole-based aggregation-induced emission (AIE)-active gold(I) complex: Persistent room-temperature phosphorescence, reversible mechanochromism and vapochromism characteristics. *Dyes Pigm.* **2017**, *143*, 409–415. [CrossRef]
18. Taranekar, P.; Fulghum, T.; Patton, D.; Ponnapati, R.; Clyde, G.; Advincula, R. Investigating carbazole jacketed precursor dendrimers: Sonochemical synthesis, characterization and electrochemical cross-linking properties. *J. Am. Chem. Soc.* **2007**, *129*, 12537–12548. [CrossRef] [PubMed]
19. Sheldrick, G.M. Crystal structure refinement with SHELXL. *Acta Crystallogr. C* **2015**, *71*, 3–8. [CrossRef] [PubMed]
20. Sheldrick, G.M. A short history of SHELX. *Acta Crystallogr. A* **2008**, *64*, 112–122. [CrossRef] [PubMed]
21. Olmstead, M.M.; Jiang, F.; Attar, S.; Balch, A.L. Alteration of the aurophilic interactions in trimeric gold(I) compounds through charge transfer. Behavior of solvoluminescent Au$_3$(MeN=COMe)$_3$ in the presence of electron acceptors. *J. Am. Chem. Soc.* **2001**, *123*, 3260–3267. [CrossRef]
22. Fujisawa, K.; Okuda, Y.; Izumi, Y.; Nagamatsu, A.; Rokusha, Y.; Sadaike, Y.; Tsutsumi, O. Reversible thermal-mode control of luminescence from liquid-crystalline gold(I) complexes. *J. Mater. Chem. C* **2014**, *2*, 3549–3555. [CrossRef]
23. Kuroda, Y.; Nakamura, S.; Srinivas, K.; Sathyanarayana, A.; Prabusankar, G.; Hisano, K.; Tsutsumi, O. Thermochemically stable liquid-crystalline gold(I) complexes showing enhanced room temperature phosphorescence. *Crystals* **2019**, *9*, 227. [CrossRef]
24. Yamada, S.; Miyano, K.; Konno, T.; Agou, T.; Kubota, T.; Hosokai, T. Fluorine-containing bistolanes as light-emitting liquid crystalline molecules. *Org. Biomol. Chem.* **2017**, *15*, 5949–5958. [CrossRef]
25. Jamain, Z.; Omar, N.F.; Khairuddean, M. Synthesis and determination of thermotropic liquid crystalline behavior of cinnamaldehyde-based molecules with two Schiff base linking units. *Molecules* **2020**, *25*, 3780. [CrossRef]
26. Hu, G.; Kitney, S.P.; Liu, Y.; Zhang, K. Synthesis and mesomorphic behavior of novel (bisthiophene)benzene carbazole nematic liquid crystals. *Mol. Cryst. Liq. Cryst.* **2021**, *723*, 81–92. [CrossRef]
27. Ge, Z.; Hayakawa, T.; Ando, S.; Ueda, M.; Akiike, T.; Miyamoto, H.; Kajita, T.; Kakimoto, M. Spin-coated highly efficient phosphorescent organic light-emitting diodes based on bipolar triphenylamine-benzimidazole derivatives. *Adv. Funct. Mater.* **2008**, *18*, 584–590. [CrossRef]

28. Tang, M.-C.; Tsang, D.P.-K.; Wong, Y.-C.; Chan, M.-Y.; Wong, K.M.-C.; Yam, V.W.-W. Bipolar gold(III) complexes for solution-processable organic light-emitting devices with a small efficiency roll-off. *J. Am. Chem. Soc.* **2014**, *136*, 17861–17868. [CrossRef] [PubMed]
29. He, X.; Lam, W.H.; Zhu, N.; Yam, V.W.-W. Design and synthesis of calixarene-based bis-alkynyl-bridged dinuclear Au^I isonitrile complexes as luminescent ion probes by modulation of Au–Au interactions. *Chem. Eur. J.* **2009**, *15*, 8842–8851. [CrossRef] [PubMed]
30. Xu, P.; Qiu, Q.; Ye, X.; Wei, M.; Xi, W.; Feng, H.; Qian, Z. Halogenated tetraphenylethene with enhanced aggregation-induced emission: An anomalous anti-heavy-atom effect and self-reversible mechanochromism. *Chem. Commun.* **2019**, *55*, 14938–14941. [CrossRef] [PubMed]
31. Bi, X.; Shi, Y.; Peng, T.; Yue, S.; Wang, F.; Zheng, L.; Cao, Q.-E. Multi-stimuli responsive and multicolor adjustable pure organic room temperature fluorescence-phosphorescent dual-emission materials. *Adv. Funct. Mater.* **2021**, *31*, 2101312. [CrossRef]
32. Favereau, L.; Quinton, C.; Poriel, C.; Roisnel, T.; Jacquemin, D.; Crassous, J. Persistent organic room-temperature phosphorescence in cyclohexane-*trans*-1,2-bisphthalimide derivatives: The dramatic impact of heterochiral vs. homochiral interactions. *J. Phys. Chem. Lett.* **2020**, *11*, 6426–6434. [CrossRef] [PubMed]
33. Tsutsumi, O.; Tamaru, M.; Nakasato, H.; Shimai, S.; Panthai, S.; Kuroda, Y.; Yamaguchi, K.; Fujisawa, K.; Hisano, K. Highly efficient aggregation-induced room-temperature phosphorescence with extremely large Stokes shift emitted from trinuclear gold(I) complex crystals. *Molecules* **2019**, *24*, 4606. [CrossRef]
34. Ando, A.; Ozaki, K.; Shiina, U.; Nagao, E.; Hisano, K.; Kamada, K.; Tsutsumi, O. Aggregation-enhanced direct S_0–T_n transitions and room-temperature phosphorescence in gold(I)-complex single crystals. *Aggregate* **2022**, *3*, e125. [CrossRef]

Article

Chiral π-Conjugated Liquid Crystals: Impacts of Ethynyl Linker and Bilateral Symmetry on the Molecular Packing and Functions

Atsushi Seki *, Kazuki Shimizu and Ken'ichi Aoki

Department of Chemistry, Faculty of Science Division II, Tokyo University of Science, 1-3 Kagurazaka, Shinjuku-ku, Tokyo 162-8601, Japan
* Correspondence: a_seki_3@rs.tus.ac.jp; Tel.: +81-3-3260-4271

Abstract: Recently, various chiral aromatic compounds, including chiral π-conjugated liquid crystals, have been developed for their unique photofunctions. One of the typical photofunctions is the bulk photovoltaic effect of ferroelectric π-conjugated liquid crystals, which integrates a polar environment based on molecular chirality with an extended π-conjugation system. Tuning the spectral properties and molecular packing is essential for improving the optical functions of the chiral π-conjugated liquid crystals. Herein, we examined the effects of an ethynyl linker and bilateral symmetry on the liquid-crystalline (LC) properties and π-conjugated system through detailed characterization via polarizing optical microscopy, differential scanning calorimetry, and X-ray diffraction analysis. The spreading of the π-conjugated system was evaluated using UV–vis absorption and photoluminescence spectroscopy. Bilateral symmetry affects the LC and photoluminescent properties. Hetero-substitution with a sparse ethynyl linker likely allows the formation of an interdigitated smectic LC structure. Because the molecular packing and photophysical properties can affect the photo- and electrical functions, we believe this study can promote the molecular design of novel functional π-conjugated materials, such as chiral ferroelectric π-conjugated liquid crystals, exhibiting the bulk photovoltaic effect.

Keywords: molecular chirality; π-conjugated compound; liquid crystal

1. Introduction

Molecular chirality can induce the formation of hierarchical suprastructures, which acts as platforms for biological, pharmacological, chemical, and physical functions [1–7]. A broken-symmetry structure is a self-organized structure that reflects the molecular chirality. Symmetry reduction leads to the stabilization of polar structures. Thus, ferroelectricity can be observed in such chiral suprastructures [8–10]. Another representative chiral supramolecular system is the helical self-assembly. The absolute configuration of the chiral molecules labeled (R) or (S) reflects their helical structure and axis. Because the helical conformation due to inherent molecular chirality is known to contribute to various functionalities of self-assembled materials, chiral materials, including chiral polymers [6,11], chiral supramolecular polymers [6,12], and chiral liquid crystals (CLCs) [2,13–15], have been extensively developed and studied. In particular, CLCs show sensitive responses to external stimuli, such as temperature and electric fields, because of their dynamic nature [14,15]. In recent years, we have focused on chiral smectic liquid crystals resulting from introducing molecular chirality into smectic liquid crystal systems [16–19]. From both basic scientific and engineering standpoints, the most important and beneficial chiral smectic liquid crystal is a ferroelectric liquid-crystalline material that exhibits a chiral smectic C (SmC*) phase. In the neutral SmC* phase, the CLC molecules form a tilted-layer structure with helical twisting of the molecular axis along the normal layer. When an electric field is applied to planar-aligned CLC molecules with a transverse dipole moment in the SmC* phase, molecular reorientation should occur to unwind helical structures. Because molecular chirality can stabilize the polar structure owing to the reduction of structural symmetry, ferroelectric properties are often observed in the SmC* phase. Conventional CLCs have been investigated for their applications such as in optical sensors [20] and high-speed liquid-crystal displays [21]. While conventional CLCs are generally insulators, chiral π-conjugated liquid crystals have the potential to be unique photofunctional materials [16–19,22]. Many π-conjugated liquid crystals have been synthesized, and their electronic charge carrier transport properties have been explored as liquid-crystalline (LC) semiconductors [23–28]. LC materials have some advantages such as improved solubility in common organic solvents, the control of molecular orientation, and the formation of uniform thin films against inorganic semiconductors. Therefore, LC semiconductors have been frequently used as active materials in optoelectronic devices such as bulk heterojunction organic photovoltaic devices [29,30], organic light-emitting diodes [31–33], and organic thin-film transistors [34–36].

This study aimed to develop chiral π-conjugated liquid crystals for novel optoelectronic materials. Recently, exciting applications of chiral π-conjugated liquid crystals have been reported. For instance, Funahashi et al. developed electric-field-responsive CLCs that exhibited an SmC* phase. As terthiophene-based CLCs show ferroelectricity and photoconductivity in the SmC* phase, the combined effect of spontaneous polarization and carrier transport results in a bulk photovoltaic effect in the LC phase [37]. The bulk photovoltaic effect based on molecular chirality is a newly discovered type of ferroelectric photovoltaic (FePV) effect, which is classified as one of the bandgap-independent photovoltaic effects [38–41]. As the FePV effect shows unique characteristics, such as ultrafast spontaneous photocurrent [42], low noise current [43], and no dissipation [44], the anomalous photovoltaic effect in ferroelectrics is evidently different from conventional photovoltaic effects based on p-n junctions [40,41]. Therefore, the FePV effect has attracted considerable attention in material chemistry and physics. While the FePV effect in ferroelectric ceramics has been widely investigated for several decades [38–44], reports on the FePV effect in organic materials other than the FePV effect of CLCs [16–19,22] are still limited [45–47]. The FePV effect for organic materials is essential for developing novel high-performance organic photoelectronic devices, including organic photovoltaic cells and organic photodetectors [47]. In fact, the FePV effect with a high open-circuit voltage of over 1 V was recently achieved by using CLCs doped with a fullerene derivative [48]. The exploration of CLCs, which are candidates for the active materials of the FePV effect, has only begun and is still important. In particular, tuning the light absorption property is a significant factor in realizing a large short-circuit current density, resulting in efficient charge carrier generation. The most common approach for tuning spectroscopic properties is expanding the π-conjugated systems, such as by introducing an ethynyl linker.

In this study, we examined the influence of the ethynyl linker between oligothiophene and chiral fluorophenyl units on the LC and photophysical properties. In addition, the impact of bilateral symmetry of the chiral compounds upon those properties were studied. We synthesized three chiral π-conjugated compounds, (R)-1, (R)-2, and (R)-3 (Figure 1). Molecular packing in the smectic LC phase and its spectroscopic properties were also investigated.

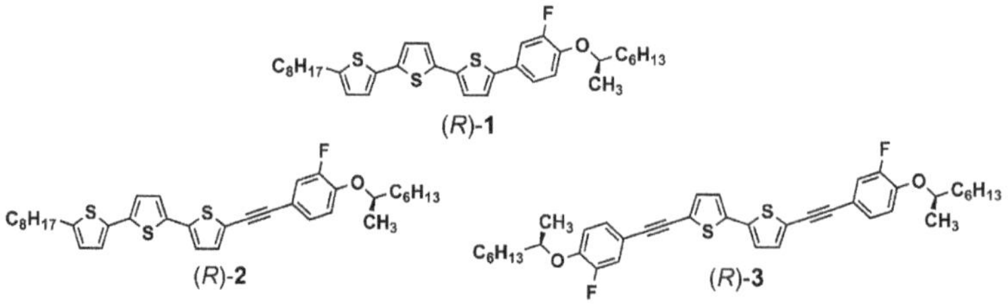

Figure 1. Chemical structures of the chiral π-conjugated compounds (R)-1, (R)-2, and (R)-3.

2. Materials and Methods

2.1. General Procedures and Materials

All reagents were purchased from Sigma-Aldrich Japan (Tokyo, Japan), Tokyo Chemical Industry Co., Ltd.(Tokyo, Japan), Kanto Chemicals (Tokyo, Japan), and FUJIFILM Wako Pure Chemicals (Osaka, Japan) and were used without further purification. All the reactions were performed under an argon atmosphere in a well-dried flask equipped with a magnetic stirring bar. The synthetic scheme for the target compounds is described in the next section (Section 2.2. Synthesis). The details of synthetic conditions are described in the attached Supplementary Files, Section S1. All ^{1}H and ^{13}C NMR spectra were recorded on a Bruker (Osaka, Japan) Biospin AVANCE NEO 400 spectrometer in CDCl$_3$ (400 MHz for ^{1}H NMR spectra, 100 MHz for ^{13}C NMR spectra). All chemical shifts (δ) in the ^{1}H and ^{13}C NMR spectra are quoted in ppm using tetramethylsilane (δ = 0.00) as the internal standard (0.03 vol%). High-resolution mass spectrometry (HRMS) measurements were carried out by electrospray ionization using a SCIEX (Tokyo, Japan) X500R QTOF spectrometer. Elemental analysis was entrusted to A-Rabbit-Science Japan Co., Ltd. (Kanagawa, Japan).

2.2. Synthesis

The chiral π-conjugated compounds (R)-**1**, (R)-**2**, and (R)-**3** were synthesized according to the procedures shown in Scheme 1. All compounds were synthesized via Pd-catalyzed C-C coupling reactions. The chiral starting material (S)-2-octanol was purchased from Tokyo Chemical Industry Co., Ltd. (Specification value: chemical purity ≥ 98.0%, optical purity ≥ 98.0%ee). Compounds **4**, (R)-**5**, **6**, (R)-**7** and **8** were synthesized with reference to literatures [17,19,49–51]. The chiral compound (R)-**5** was synthesized via the Suzuki–Miyaura reaction between 2,2′-bithiophene-5-boronic acid pinacol ester and 4-bromo-2-fluoro-1-{(R)-2-octyloxy}benzene. 4-Bromo-2-fluoro-1-{(R)-2-octyloxy}benzene was synthesized via the Mitsunobu reaction between 4-bromophenol and (S)-2-octanol. It is noted that the optical purity of (S)-2-octanol is guaranteed ≥ 98.0%ee by the standard. Because the Mitsunobu reaction generally undergoes the typical S$_N$2 displacement pathway, chiral inversion must be caused [52]. Shi et al. reported that the Mitsunobu reaction using chiral alcohols exhibiting high enantiomeric excess (> 90%ee) with phenol derivatives afford the product with high optical purity (> 90%ee) [53]. Based on these findings, various chiral liquid crystals have been synthesized from (S)-2-octanol or (R)-2-octanol via the several reaction steps including Mitsunobu reaction and C-C cross-coupling reactions [54–57]. ^{1}H-, ^{13}C NMR and HRMS spectra for the target compounds (R)-**1**, (R)-**2**, and (R)-**3** are shown in the ESI, Sections S2 and S3.

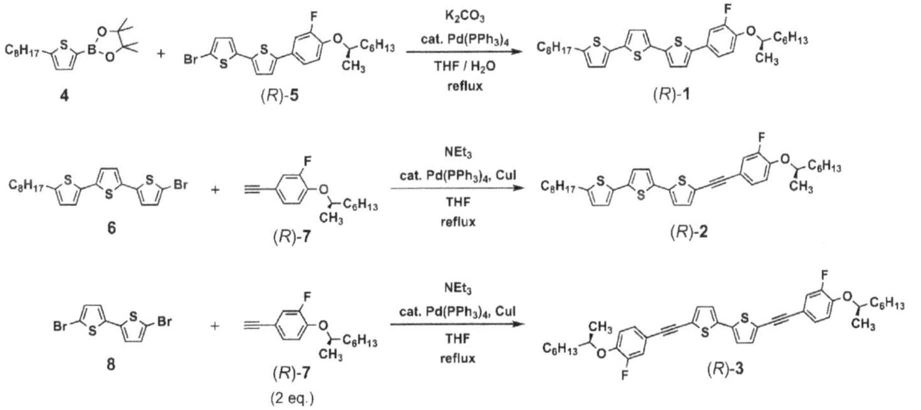

Scheme 1. Synthesis routes of compounds (R)-**1**, (R)-**2**, and (R)-**3**.

2.2.1. Characterization of (R)-1

5-Octyl-5″-{3-fluoro-4-[(R)-2-octyloxy]phenyl}-2,2′:5′,2″-terthiophene: (R)-1

^{1}H NMR (400 MHz, CDCl$_3$): δ [ppm] = 7.30 (dd, 1H, J = 12.2, 2.2 Hz), 7.25 (ddd, 1H, J = 8.4, 2.4, 1.2 Hz), 7.10 (d, 1H, J = 3.6 Hz), 7.08 (d, 1H, J = 3.6 Hz), 7.05 (d, 1H, J = 3.6 Hz), 6.99 (d, 1H, J = 4.0 Hz), 6.98 (d, 1H, J = 3.2 Hz), 6.95 (t, 1H, J = 8.6 Hz), 6.68 (d, 1H, J = 3.6 Hz), 4.37 (sextet, 1H, J = 6.0 Hz), 2.79 (t, 2H, J = 7.4 Hz), 1.86–1.55 (m, 4H), 1.51–1.20 (m, 18H), 1.33 (d, 3H, J = 6.0 Hz), 0.89 (t, 6H, J = 7.0 Hz); ^{13}C NMR (100 MHz, CDCl$_3$): δ [ppm] = 153.8 (d, J = 244.2 Hz), 145.8 (d, J = 10.9 Hz), 145.7, 141.8 (d, J = 2.2 Hz), 136.9, 136.3, 135.4, 134.4, 127.8 (d, J = 7.2 Hz), 124.9, 124.3, 124.1, 123.5 (d, J = 15.9 Hz), 123.4, 121.4 (d, J = 3.6 Hz), 117.9 (d, J = 2.7 Hz), 113.8, 113.6, 76.6, 36.5, 31.9, 31.8, 31.6, 30.2, 29.3, 29.3, 29.2, 29.1, 25.4, 22.7, 22.6, 19.8, 14.1, 14.1; HRMS (ESI): molecular weight: 582.8954 (C$_{34}$H$_{43}$FOS$_3$); m/z calculated for [C$_{34}$H$_{43}$FOS$_3$]$^+$: 582.2455 ([M]$^+$); found: 582.2456; elemental analysis (%) calculated for C$_{34}$H$_{43}$FOS$_3$: C 70.06, H 7.44, F 3.26, O 2.74, S 16.50; found: C 69.81, H 7.28.

2.2.2. Characterization of (R)-2

5-Octyl-5″-({3-fluoro-4-[(R)-2-octyloxy]phenyl}ethynyl)-2,2′:5′,2″-terthiophene: (R)-2

^{1}H NMR (400 MHz, CDCl$_3$): δ [ppm] = 7.25–7.18 (m, 2H), 7.14 (d, 2H, J = 3.6 Hz), 7.07 (d, 1H, J = 4.0 Hz), 7.03 (d, 1H, J = 4.0 Hz), 6.99 (d, 1H, J = 4.0 Hz), 6.99 (d, 1H, J = 3.6 Hz), 6.91 (t, 1H, J = 8.8 Hz), 4.39 (sextet, 1H, J = 6.0 Hz), 2.79 (t, 2H, J = 7.6 Hz), 1.85–1.55 (m, 4H), 1.52–1.22 (m, 18H), 1.33 (d, 3H, J = 6.4 Hz), 0.88 (t, 6H, J = 6.8 Hz); ^{13}C NMR (100 MHz, CDCl$_3$): δ [ppm] = 152.9 (d, J = 244.8 Hz), 147.0 (d, J = 10.9 Hz), 146.0, 138.7, 137.6, 134.7, 134.2, 132.7, 127.9 (d, J = 3.6 Hz), 124.9, 124.8, 123.6 (d, J = 3.2 Hz), 123.2, 121.7, 119.4, 119.2, 116.7 (d, J = 2.4 Hz), 115.3 (d, J = 8.3 Hz), 93.2, 82.0, 76.3, 36.4, 31.9, 31.8, 31.6, 30.2, 29.3, 29.2, 29.1, 25.4, 22.7, 22.6, 19.8, 14.1, 14.1; HRMS (ESI): molecular weight: 606.9174 (C$_{36}$H$_{43}$FOS$_3$); m/z calculated for [C$_{36}$H$_{43}$FOS$_3$]$^+$: 606.2455 ([M]$^+$); found: 606.2453; elemental analysis (%) calculated for C$_{36}$H$_{43}$FOS$_3$: C 71.24, H 7.14, F 3.13, O 2.64, S 15.85; found: C 71.25, H 7.16.

2.2.3. Characterization of (R)-3

5,5′-Bis({3-fluoro-4-[(R)-2-octyloxy]phenyl}ethynyl)-2,2′-bithiophene: (R)-3

^{1}H NMR (400 MHz, CDCl$_3$): δ [ppm] = 7.25–7.19 (m, 4H), 7.15 (d, 2H, J = 3.6 Hz), 7.07 (d, 2H, J = 4.0 Hz), 6.92 (t, 2H, J = 8.6 Hz), 4.40 (sextet, 2H, J = 6.2 Hz), 1.85–1.72 (m, 2H), 1.67–1.55 (m, 2H), 1.52–1.24 (m, 16H), 1.33 (d, 6H, J = 6.4 Hz), 0.88 (t, 6H, J = 6.8 Hz); ^{13}C NMR (100 MHz, CDCl$_3$): δ [ppm] = 152.9 (d, J = 245.4 Hz), 147.1 (d, J = 10.8 Hz), 138.0, 132.7, 128.0 (d, J = 3.0 Hz), 124.0, 122.5, 119.3 (d, J = 20.2 Hz), 116.7 (d, J = 2.4 Hz), 115.2 (d, J = 8.5 Hz), 93.5 (d, J = 2.8 Hz), 81.8, 76.3, 36.4, 31.8, 29.2, 25.4, 22.6, 19.8, 14.1; HRMS (ESI): molecular weight: 658.9068 (C$_{40}$H$_{44}$F$_2$O$_2$S$_2$); m/z calculated for [C$_{40}$H$_{45}$F$_2$O$_2$S$_2$]$^+$: 659.2824 ([M+H]$^+$); found: 659.2828; elemental analysis (%) calculated for C$_{40}$H$_{44}$F$_2$O$_2$S$_2$: C 72.91, H 6.73, F 5.77, O 4.86, S 9.73; found: C 72.99, H 6.86.

2.3. Characterization of LC Properties

The LC properties of chiral π-conjugated compounds were characterized using differential scanning calorimetry (DSC), polarizing optical microscopy (POM), and X-ray diffraction (XRD). DSC measurements were conducted using a SHIMADZU (Kyoto, Japan) DSC-60 system equipped with a liquid nitrogen auto-cooling system (TAC-60L). Approximately 2–3 mg of each sample was sealed in an aluminum pan. The optical texture was observed using a polarizing optical microscope (Olympus BH2, Olympus Corporation, Tokyo, Japan) equipped with a digital camera (AS ONE HDCE-X1 (AS ONE Corporation, Osaka, Japan) and a temperature control system (METTLER TOLEDO FP90 and FP82HT). Indium tin oxide (ITO) sandwich cells filled with chiral π-conjugated compounds were used for POM observations. Empty ITO sandwich cells (KSSO-02/A311P1NSS05, cell gap: 2 μm) were purchased from EHC Corporation (Tokyo, Japan). The ITO surface without a polyimide was rubbed to assist in the planar orientation of the smectic phases. The scan rate of DSC measurements and POM observations was 10 °C min^{-1}. XRD measurements were performed using a Rigaku RINT-2500 (Ni-filtered Cu Kα radiation, Rigaku Corporation,

Tokyo, Japan) equipped with a custom-made thermal control system composed of a silicone rubber heater, thermocouple sensor, and PID-type thermal controller (AS ONE TJA-550).

2.4. Characterization of Spectroscopic Properties

UV–vis absorption spectra were recorded using a JASCO (Tokyo, Japan) V-650 spectrometer. UV–vis absorption spectra were measured using a pair of quartz cells (cell gap: 1 cm). The photoluminescence (PL) emission spectra were recorded using a SHIMADZU (Kyoto, Japan) RF-6000 spectrometer. Emission spectra were measured using a pair of quartz cells (cell gap: 1 cm).

3. Results and Discussion
3.1. Liquid-Crystalline Properties
3.1.1. Polarizing Optical Microscopy

In the POM observation of chiral phenylterthiophene derivative (R)-**1**, a broken fan-like texture with stripes was observed in the area where the sample was sandwiched between two ITO electrodes, at approximately 140 °C upon cooling from the isotropic liquid (IL) state (Figure 2a). The broken fan-like domains suggest the formation of an LC tilted-layer structure. Furthermore, the stripe pattern in each fan-shaped domain should be derived from the disclination. Therefore, the characteristic optical textures indicated the appearance of a chiral smectic C (SmC*) phase with a nonpolar helical structure. When the sample was cooled to approximately 130 °C, the polarized optical texture was transformed, and tile-like domains were observed in the homeotropic domains of the SmC* phase between the two glass substrates (Figure 2b). This change in texture corresponds to a phase transition from SmC* to ordered smectic phases. Upon further cooling to room temperature, the domain shapes were maintained without drastic textural changes (Figure 2c). In the POM study of compound (R)-**2**, an ethynyl linker introduced between the terthiophene and chiral fluorophenyl units, we observed a fan-shaped texture upon cooling from the IL phase (Figure 3a). A typical fan-shaped texture shows the appearance of a smectic LC phase. Because the color and contrast of the optical texture vary as the sample temperature of (R)-**2** decreases to approximately 110 °C (Figure 3b), the high-temperature smectic LC phase is changed to another smectic LC phase at this temperature. After cooling below 100 °C, the stripes appeared in fan-shaped domains (Figure 3c). In the polarized optical texture of (R)-**2** at 45 °C, the stripes of fan-shaped domains are more conspicuous (Figure 3d). The slight change of optical texture is probably due to rearrangements in the intralayer molecular packing. These results support that (R)-**2** exhibited several smectic LC phases. In the POM observation of compound (R)-**3** modified with chiral fluorophenyl units on both wings of the 2,2′-bithiophene core, two types of optical textures were observed during cooling from the IL phase (Figure 4a,b). However, these textures differ from the distinctive textures of LC phases. Therefore, we conclude that (R)-**3** does not exhibit LC properties.

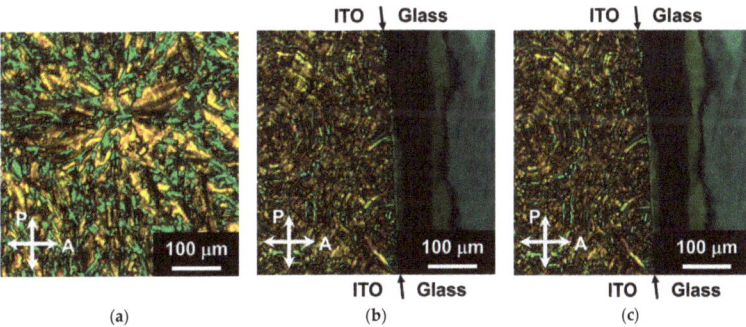

Figure 2. POM images of (R)-**1** at (**a**) 140 °C, (**b**) 95 °C, and (**c**) 40 °C. The black arrows indicate the border of ITO electrode and glass surface.

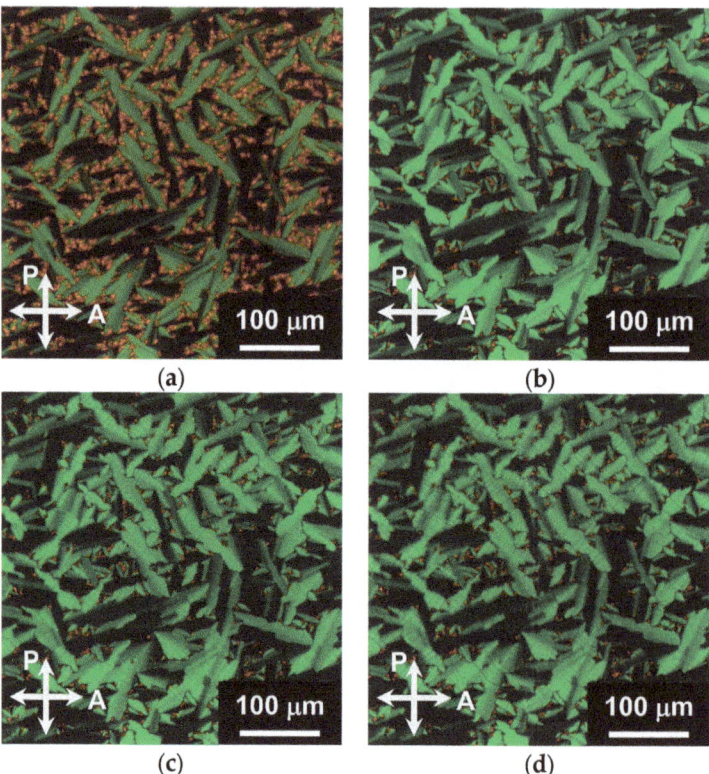

Figure 3. POM images of (R)-**2** at (**a**) 120 °C, (**b**) 105 °C, (**c**) 90 °C, and (**d**) 45 °C.

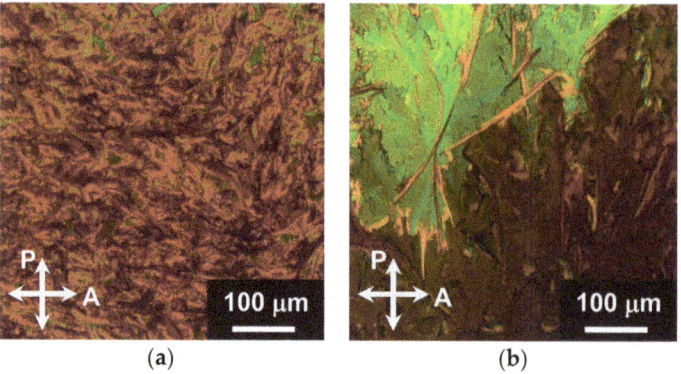

Figure 4. POM images of (R)-**3** at (**a**) 70 °C, and (**b**) 40 °C.

3.1.2. Differential Scanning Calorimetry

The DSC thermogram of (R)-**1** exhibits two distinct endothermic peaks due to first-order phase transitions during the second heating (Figure 5a). Although a crystal-LC phase transition peak is found at 63.3 °C on the first heating, no endothermic peak of crystal-LC transition is seen on subsequent heating scans. Similar phase transition behaviors are found in analogous phenylterthiophene derivatives [17,37]. The inconsistency in the number of first-order phase transition peaks between first cooling and second heating scans suggests the existence of a monotropic metastable mesophase during cooling. When we consider the results of POM and DSC studies of (R)-**1**, the exothermic peak at 144.2 °C during cooling indicates the IL–SmC* phase transition. The following peak observed at 131.3 °C corresponds to the transition from SmC* to metastable ordered smectic phases. The metastable smectic phase transforms to a more stable ordered smectic phase at 66.5 °C during cooling. On the second heating scan, the endothermic peak of the ordered smectic-SmC* phase transition is observed at 132.5 °C, and the SmC*-IL phase transition follows at 144.8 °C. In the DSC thermogram of (R)-**2**, three peaks are observed during the second heating (Figure 5b). These peaks correspond to the two LC–LC phase transitions and an LC–IL phase transition based on the results of the POM study. The middle LC phase changes to the low-ordered smectic phase at 111.5 °C after the first LC–LC phase transition occurs at 101.3 °C. The smectic LC structure and the molecular order collapses at 126.0 °C. The broad exothermic peak, at approximately 67 °C, is observed on the first cooling scan, corresponding to the transition from metastable to stable states (Figure 5b). A metastable LC phase is observed during cooling for compound (R)-**2**. Therefore, we conclude that compounds (R)-**1** and (R)-**2** exhibit similar monotropic behavior. The broad tolerance of molecular packing style in (R)-**2** should be also reflected to the complicated phase transition behaviors on the first heating process. Compound (R)-**3** also shows several phase transitions at 52.3 and 77.2 °C, as observed in the second heating scan (Figure 5c). The DSC and POM studies support that compound (R)-**3** exhibits crystalline polymorphism.

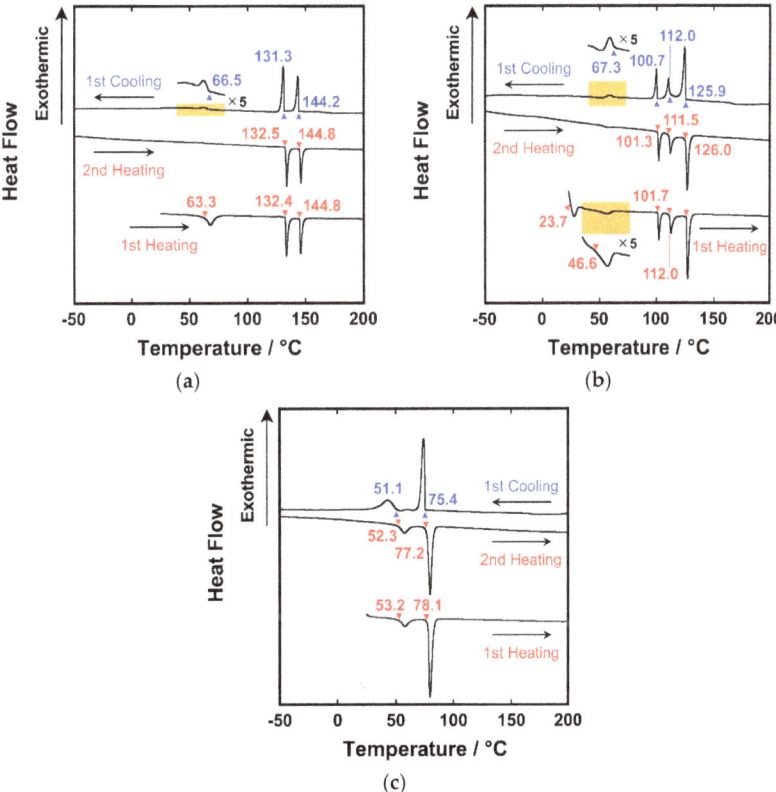

Figure 5. DSC thermograms of (**a**) (R)-**1**, (**b**) (R)-**2**, and (**c**) (R)-**3** at a scanning rate of 10 °C min^{-1}.

3.1.3. X-ray Diffraction

The variable-temperature XRD measurements were conducted for the chiral π-conjugated compounds (R)-**1**, (R)-**2**, and (R)-**3** to gain insight into their molecular packing and self-assembled structures. The XRD pattern of (R)-**1** in the SmC* phase at 139 °C (Figure 6a, upper) exhibits diffraction peaks at 2θ = 3.12°, 6.10°, 12.11°, 15.15°, and 18.21°, which correspond to diffractions from the (001), (002), (003), (004), (005), and (006) planes, respectively. Because all d-spacings estimated from these diffractions can be expressed as integer ratios, the XRD pattern also indicates that a smectic-layer structure at approximately 140 °C can be formed. The molecules in the smectic-layer structure should be tilted with respect to the normal of the layer because the layer spacing (29 Å) is shorter than the theoretical extended molecular length (35 Å) of (R)-**1** estimated by the molecular mechanics calculation (Energy minimization calculation, MM2 force field, PerkinElmer, Chem3D 18.1). The SmC* phase is observed between 144 and 131 °C upon cooling of (R)-**1**. This observation coincides with those of the preceding POM and DSC studies. In the XRD pattern, during cooling, we observed a diffraction peak with a low intensity at 95 °C in the wide-angle region (2θ = 19.30°) and several other diffraction peaks that represented from the smectic-layer structures (Figure 6a, middle). The low-intensity peak can be assigned to the (010) plane, reflecting the intralayer order. From the periodic diffraction peaks corresponding to the (001), (002), (003), (004), and (006) planes, the layer spacing is estimated to be 29 Å. Therefore, the tilt angle remains unchanged through the SmC*–LC phase transition. As no other peaks are observed in the wide-angle region, the intralayer order in the metastable smectic phase should be confined in the short range. Therefore, we consider the metastable phase at 95 °C to be an ordered chiral smectic (SmX$_1$*) phase which is probably either chiral smectic F or chiral smectic I phase [57–60]. The XRD profile of the more stable highly ordered chiral smectic (SmX$_2$*) phase at 27 °C differed from those of the SmC* and SmX$_1$* phases (Figure 6a, lower). The increase in the intralayer molecular order is indicated by a broad peak observed at 2θ = 18.64° for the (100) plane and by an increase in the relative peak intensity of the (010) diffraction. The shorter layer spacing of 29 Å and the calculated molecular length indicate that the tilted-layer structure is maintained even in the ordered smectic phase. This ordered smectic (SmX$_2$*) phase should be one of the chiral smectic G, chiral smectic J or chiral smectic H phase, as determined by the general phase transition sequence [57–61].

The XRD pattern of the LC phase of (R)-**2** at 124 °C (Figure 6b, upper) exhibits several peaks at 2θ = 2.16°, 4.36°, and 6.49°. These three peaks are attributed to the (001), (002), and (003) planes with diffractions derived from the periodicity of the smectic-layer structure. Although the extended molecular length of (R)-**2** is estimated to be 38 Å by MM2 calculations, the experimentally obtained layer spacing is 41 Å. The layer spacing is greater than the theoretical molecular length that an interdigitated layer structure can achieve. The halos observed at approximately 2θ = 12°, 20°, 24°, and 26° also confirm the interdigitated organization of (R)-**2** molecules. Because the POM textures of (R)-**2** at a comparable temperature are typical for a low-ordered smectic phase and not for a characteristic texture for a highly ordered smectic phase, the formation of a highly ordered smectic phase is uncertain. In addition, the sample of (R)-**2** shows fluidity in the LC phase. These behaviors can be observed in a low-ordered interdigitated smectic phase. While the halo at 2θ = 12° can be ascribed to the disordered aggregation of bulky chiral alkyl chains based on steric effects, the series of halos between 2θ = 18° and 2θ = 30° probably resulted from the disordered aggregation of the linear aliphatic chains and interaction between aromatic units. Thus, the appearance of several halos suggests that each of the rigid aromatic units and mobile chiral alkyl chains is segregated and gathered in different periodicities. The integrated molecules of (R)-**2** in the LC phase should be tilted with respect to the layer normal, considering the molecular packing model (Figure 7). Therefore, upon cooling, we assigned the LC phase of (R)-**2** between 126 °C and 112 °C to an interdigitated chiral smectic C (SmC$_d$*) phase [62–67]. When the XRD sample of (R)-**2** was cooled to 108 °C, the normalized intensities of the (002) and (003) diffraction peaks in the XRD profile (Figure 6b, middle) were higher than those of the same peaks in the XRD pattern of the

SmC$_d$* phase (Figure 6b, upper). The absence of sharp diffraction peaks in the wide-angle region indicates the absence of long-range intralayer order in the middle-temperature LC phase. Because the interlayer spacing undergoes a slight change of 41–42 Å via the SmC$_d$*–LC phase transition, the interdigitated layer structure is maintained. Based on these results, we believe that the middle-temperature LC phase is a chiral smectic (SmX$_{d1}$*) phase, in which the interdigitated LC structures have short-range intralayer order. In the XRD pattern of (R)-2 cooled to room temperature (34 °C), additional weak diffraction peaks were observed at 2θ = 10.6° and 19.4° (Figure 6b, bottom). These peaks originated from the (005) and (010) diffraction planes. Because the sharp (010) diffraction peak indicates growing intralayer-bond order, the room-temperature LC phase is identified as a highly ordered interdigitated chiral smectic (SmX$_{d2}$*) phase. The small difference of XRD patterns between SmX$_{d1}$* phase and SmX$_{d2}$* phase suggests the slight structural change through phase transitions via the metastable state. The POM study of (R)-2 on cooling process (Figure 3b–d) also supports this consideration. The metastable phase of (R)-2 probably appears while the rearrangement of intralayer molecular packing proceeds under the influence of spatial factors by bulky chiral unit and sparse ethynyl moiety.

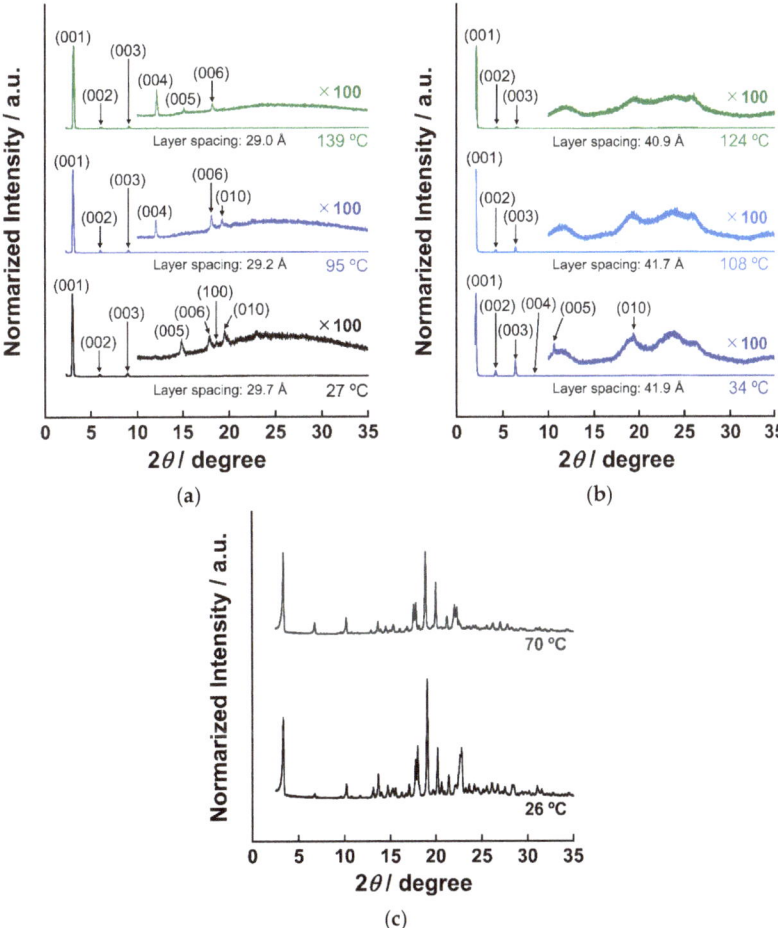

Figure 6. Variable-temperature XRD profiles of (**a**) (R)-**1**, (**b**) (R)-**2**, and (**c**) (R)-**3**.

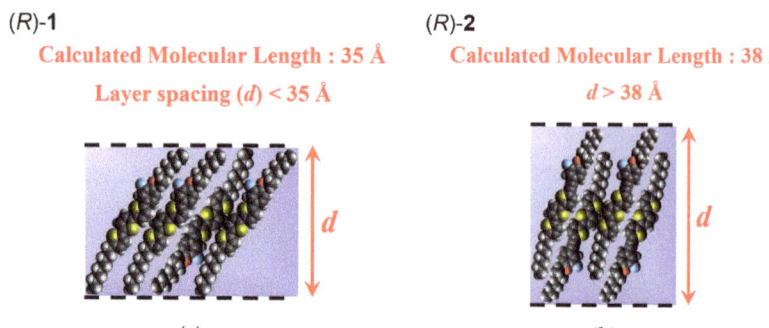

Figure 7. Schematic illustrations of the molecular packing models in the (**a**) SmC* phase of (*R*)-**1** (monolayer structure), and (**b**) SmC$_d$* phase of (*R*)-**2** (interdigitated layer structure) assumed from the XRD profiles.

Compound (*R*)-**3** exhibited complicated XRD patterns, indicating crystalline molecular packing below 74 °C (Figure 6c). Although several phase transition peaks were observed in the DSC thermogram, bis({fluorophenyl}ethynyl) bithiophene (*R*)-**3** showed crystal polymorphism and did not show any thermodynamically stable LC phase.

The phase transition behaviors of (*R*)-**1**, (*R*)-**2**, and (*R*)-**3** are summarized in Table 1. For each of LC compounds (*R*)-**1** and (*R*)-**2**, the initial crystalline precipitates for the characterization of LC properties were obtained by recrystallization. It is noted that both LC compounds (*R*)-**1** and (*R*)-**2** exhibit the crystalline–LC phase transition only in the first heating process. Once the precipitates melted to the IL phase, no crystallization occurred during the cooling process below −50 °C at a scanning rate of 10 °C min^{-1}. These results show that a bilateral asymmetric molecular structure is effective for liquid crystallinity. In addition, we consider that the interplay of bulky chiral unit and sparse ethynyl moiety prominently causes a variety of molecular packing as well as the formation of interdigitated structures.

Table 1. Phase transition behavior of (*R*)-**1**, (*R*)-**2**, and (*R*)-**3**.

Compound		Phase Transition Temperature/°C (Enthalpy/kJ mol^{-1})
(*R*)-**1**	first heating first cooling second heating	Cr 63 (−13) SmX$_2$* 132 (−13) SmC* 145 (−11) IL IL 144 (11) SmC* 131 (12) SmX$_1$* 67 (1) SmX$_2$* SmX$_2$* 133 (−13) SmC* 145 (−11) IL
(*R*)-**2**	first heating first cooling second heating	Cr 24 (−5) M 47 (−2) SmX$_{d2}$* 102 (−3) SmX$_{d1}$* 112 (−4) SmC$_d$* 126 (−9) IL IL 126 (9) SmC$_d$* 112 (3) SmX$_{d1}$* 101 (4) SmX$_m$* 67 (1) SmX$_{d2}$* SmX$_{d2}$* 101 (−3) SmX$_{d1}$* 112 (−3) SmC$_d$* 126 (−8) IL
(*R*)-**3**	first heating first cooling second heating	Cr$_2$ 53 (−9) Cr$_1$ 78 (−37) IL IL 75 (33) Cr$_1$ 51 (18) Cr$_2$ Cr$_2$ 52 (−9) Cr$_1$ 77 (−36) IL

The abbreviations Cr, Cr$_1$, Cr$_2$, IL, M, SmC*, SmC$_d$*, SmX$_1$*, SmX$_2$*, SmX$_{d1}$*, SmX$_{d2}$*, and SmX$_m$* denote crystalline, crystalline 1, crystalline 2, isotropic liquid, unidentified ordered, chiral smectic C, chiral interdigitated smectic C, unidentified ordered chiral smectic 1, unidentified ordered chiral smectic 2, unidentified interdigitated ordered smectic 1, interdigitated ordered smectic 2, and metastable interdigitated ordered smectic phases, respectively.

3.2. Spectroscopic Properties

Figure 8a shows the UV–vis absorption and PL spectra in a dilute THF solution of (*R*)-**1**, (*R*)-**2**, and (*R*)-**3**. The absorption spectrum of (*R*)-**1** in THF (10 μM) showed a quasi-unimodal absorption band corresponding to the π–π* transition of the terthiophene unit between 330 and 450 nm. The absorption maximum was 393 nm with a molar absorption coefficient of 4.0×10^4 L mol^{-1} cm^{-1}. By comparing the absorption spectra of (*R*)-**1** and

(R)-2 in THF dilute solutions, a slight shift in the absorption band of (R)-2 is observed towards the longer wavelength region. This result suggested that introducing an ethynyl linker to the mesogenic core accurately extended the effective π-conjugation length. In addition, the THF solution of compound (R)-2 showed a higher molar absorption coefficient of 4.7×10^4 L mol^{-1} cm^{-1} at the absorption maximum (λ_{abs} = 397 nm) compared to those of the solution of (R)-1. The absorption spectrum of (R)-3 in THF (10 µM) displays the π–π* transition band of the bithiophene core with an absorption maximum of 390 nm. For a dilute solution of (R)-3, the molar absorption coefficient attained 5.5×10^4 L mol^{-1} cm^{-1}. The absorption edges in the THF solutions of (R)-1, (R)-2, and (R)-3 are 446, 450, and 455 nm, respectively (Figure 8b). Because the order of the absorption maxima and edges reflects the π-conjugation length, (R)-2 should have the longest effective π-conjugation length among the three compounds. Each fluorescence spectrum indicates well-resolved vibrational structures (Figure 8a). The maximum PL intensity of (R)-3 in the THF solution was more than twice those of (R)-1 and (R)-2. This result indicates that modifying phenylethynyl units on both wings of the 2,2'-bithiophene core enhances fluorescence emission. The luminescence enhancement appears to result from the suppression of thermal relaxation, and an increase in the oscillator strength is observed. In the case of compound (R)-3, introducing an ethynyl linker may reduce steric interactions and extend the π-conjugation length [68]. Table 2 lists the spectroscopic parameters. The disubstituted compound (R)-3 showed the slightest Stokes shift among the three compounds, and phenylethynyl terthiophene (R)-2 displayed the most significant Stokes shift. According to previous reports, we considered that the difference in Stokes shift originated from conformational changes rather than solvent effects [68]. The difference in Stokes shift implies a variation in the conformational change between the ground and excited states. The modification of the chiral phenylethynyl units showed different effects on excitation and emission in (R)-3 and (R)-2. Because interorbital electronic interactions are sensitive to modifications, the principal cause of the difference in spectroscopic properties of (R)-2 and (R)-3 seems to be left-right asymmetrical hetero-substitution. The lack of a drastic increase in the PL intensity of (R)-2 also supports this hypothesis.

Table 2. Spectroscopic properties of (R)-1, (R)-2, and (R)-3 in a dilute THF solution (10 µM).

Compound	λ_{abs}/nm (Molar Absorption Coefficient/L mol^{-1} cm^{-1})	λ_{em}/nm	Stokes Shift/eV
(R)-1	393 (4.0×10^4)	451, 478 [a]	0.406
(R)-2	397 (4.7×10^4)	458, 483 [b]	0.416
(R)-3	390 (5.5×10^4)	445, 471 [c]	0.393

[a] Excitation wavelength of 393 nm. [b] Excitation wavelength: 397 nm. [c] The excitation wavelength was 390 nm.

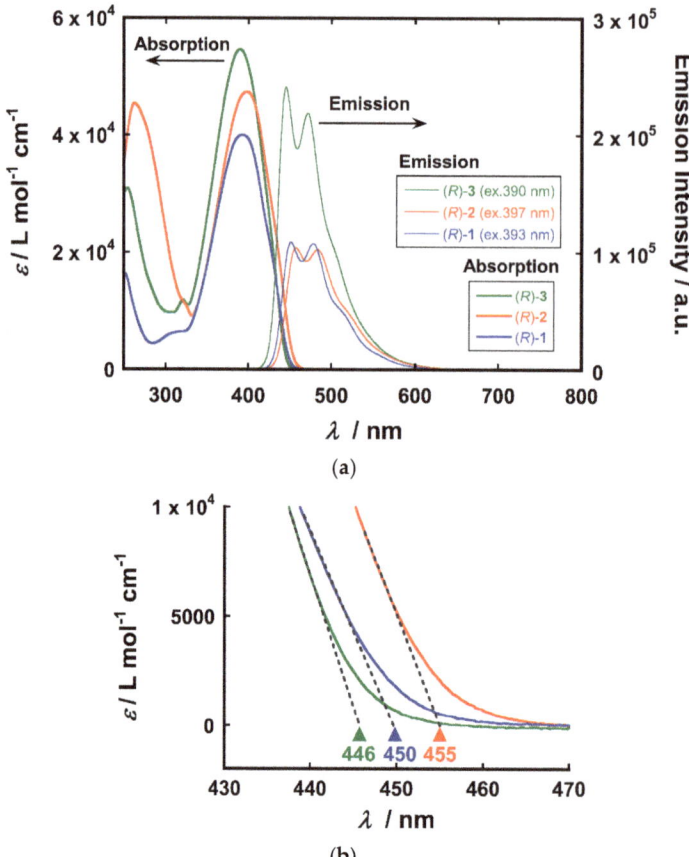

Figure 8. (a) UV–vis absorption and photoluminescent spectra in THF dilute solution (10 μM) of (R)-**1**, (R)-**2**, and (R)-**3**. (b) The magnified UV–vis absorption spectra in (a). Each triangle marks in the inset of absorption spectra depicting the absorption edge of (R)-**1** (blue), (R)-**2** (red), and (R)-**3** (green).

4. Conclusions

We synthesized three chiral oligo-thiophene derivatives, (R)-**1**, (R)-**2**, and (R)-**3**. While compounds (R)-**1** and (R)-**2** exhibited chiral smectic LC phases, the disubstituted bithiophene analog (R)-**3** showed only crystalline polymorphism. This outcome suggests that the bilateral symmetry hinders liquid-crystallinity. In other words, molecular structuring with left-right asymmetry promotes the formation of LC structures. The ethynyl-inserted mono-substituted compound (R)-**2** forms LC interdigitated layer structures due to the steric effect of the spatially sparse ethynyl linker and the bulky chiral moiety. In addition, the interplay of both units could effectively contribute to the formation of various smectic LC structures. The UV–vis absorption and PL spectra in a dilute THF solution indicate that (R)-**2** has a more expansive π-conjugation system than (R)-**1** because of the introduction of the ethynyl linker. The ethynyl linker also contributed to an increase in the molar absorption coefficient in the visible region. Because the molecular packing and photophysical properties affect the photoconductive properties, introducing an ethynyl linker in the central π-conjugated core causes drastic changes in the performance of organic optoelectronic devices. We believe our results can help in the molecular design of novel functional chiral π-conjugated liquid crystals, including ferroelectric π-conjugated liquid crystals that exhibit the FePV effect.

Supplementary Materials: The following supporting information can be downloaded from https://www.mdpi.com/article/10.3390/cryst12091278/s1, Section S1. Synthetic procedure; Section S2. ^{1}H and ^{13}C NMR spectra (Figure S1. ^{1}H NMR spectrum of (R)-**1**; Figure S2. ^{13}C NMR spectrum of (R)-**1**; Figure S3. ^{1}H NMR spectrum of (R)-**2**; Figure S4. ^{13}C NMR spectrum of (R)-**2**; Figure S5. ^{1}H NMR spectrum of (R)-**3**; Figure S6. ^{13}C NMR spectrum of (R)-**3**); Section S3. High-resolution electrospray ionization (ESI) mass spectra (Figure S7. High-resolution ESI mass spectrum of (R)-**1**; Figure S8. High-resolution ESI mass spectrum of (R)-**2**; Figure S9. The high-resolution ESI mass spectrum of (R)-**3**).

Author Contributions: Conceptualization, A.S.; methodology, A.S.; validation, A.S., K.S. and K.A.; formal analysis, A.S. and K.S.; investigation, A.S. and K.S.; resources, A.S. and K.A.; data curation, A.S. and K.S.; writing—original draft preparation, A.S.; writing—review and editing, A.S. and K.S.; visualization, A.S. and K.S.; supervision, A.S. and K.A.; project administration, A.S.; funding acquisition, A.S. and K.A. All authors have read and agreed to the published version of the manuscript.

Funding: This study was financially supported by a research fund from the Tokyo University of Science for A.S. and K.A. and a research grant from the Amano Institute of Technology, Japan for A.S.

Institutional Review Board Statement: Not applicable.

Informed Consent Statement: Not applicable.

Data Availability Statement: Not applicable.

Acknowledgments: The authors thank Khoa V. Le and T. Sasaki at the Tokyo University of Science for their help with POM observations. We also greatly appreciate the help of HRMS measurements by Y. Yoshimura at the Tokyo University of Science.

Conflicts of Interest: The authors declare no conflict of interest.

References

1. Kasprzyk-Hordern, B. Pharmacologically active compounds in the environment and their chirality. *Chem. Soc. Rev.* **2010**, *39*, 4466–4503. [CrossRef] [PubMed]
2. Bisoyi, H.K.; Li, Q. Light-Directing Chiral Liquid Crystal Nanostructures: From 1D to 3D. *Acc. Chem. Res.* **2014**, *47*, 3184–3195. [CrossRef] [PubMed]
3. Pescitelli, G.; Di Bari, L.; Berova, N. Application of electronic circular dichroism in the study of supramolecular systems. *Chem. Soc. Rev.* **2014**, *43*, 5211–5233. [CrossRef] [PubMed]
4. Zhang, L.; Qin, L.; Wang, X.; Cao, H.; Liu, M. Supramolecular Chirality in Self-Assembled Soft Materials: Regulation of Chiral Nanostructures and Chiral Functions. *Adv. Mater.* **2014**, *26*, 6959–6964. [CrossRef] [PubMed]
5. Zhang, L.; Wang, T.; Shen, Z.; Liu, M. Chiral Nanoarchitectonics: Towards the Design, Self-Assembly, and Function of Nanoscale Chiral Twists and Helices. *Adv. Mater.* **2016**, *28*, 1044–1059. [CrossRef]
6. Yashima, E.; Ousaka, N.; Taura, D.; Shimomura, K.; Ikai, T.; Maeda, K. Supramolecular Helical Systems: Helical Assemblies of Small Molecules, Foldamers, and Polymers with Chiral Amplification and Their Functions. *Chem. Rev.* **2016**, *116*, 13752–13990. [CrossRef]
7. Evers, F.; Aharony, A.; Bar-Gill, N.; Entin-Wohlman, O.; Hedegaard, P.; Hod, O.; Jelinek, P.; Kamieniarz, G.; Lemeshko, M.; Michaeli, K.; et al. Theory of Chirality Induced Spin Selectivity: Progress and Challenges. *Adv. Mater.* **2022**, *34*, 2106629. [CrossRef]
8. Meyer, R.B.; Libert, L.; Strzelecki, L.; Keller, P. Ferroelectric liquid crystals. *J. Phys.* **1975**, *36*, L69–L71. [CrossRef]
9. Young, C.Y.; Pindak, R.; Clark, N.A.; Meyer, R.B. Light-Scattering Study of Two-Dimensional Molecular-Orientation Fluctuations in a Freely Suspended Ferroelectric Liquid-Crystal Film. *Phys. Rev. Lett.* **1978**, *40*, 773–776. [CrossRef]
10. Gao, W.; Zhang, Z.; Li, P.F.; Tang, Y.Y.; Xiong, R.G.; Yuan, G.; Ren, S. Chiral Molecular Ferroelectrics with Polarized Optical Effect and Electroresistive Switching. *ACS Nano* **2017**, *11*, 11739–11745. [CrossRef]
11. Scanga, R.A.; Reuther, J.F. Helical polymer self-assembly and chiral nanostructure formation. *Polym. Chem.* **2021**, *12*, 1857–1897. [CrossRef]
12. Dorca, Y.; Greciano, E.E.; Valera, J.S.; Gomez, R.; Sanchez, L. Hierarchy of Asymmetry in Chiral Supramolecular Polymers: Toward Functional, Helical Supramolecular Structures. *Chem. Eur. J.* **2019**, *25*, 5848–5864. [CrossRef] [PubMed]
13. Dierking, I. Chiral Liquid Crystals: Structures, Phases, Effects. *Symmetry* **2014**, *6*, 444–472. [CrossRef]
14. Kitzerow, H.-S.; Bahr, C. (Eds.) *Chirality in Liquid Crystals*, 1st ed.; Springer: New York, NY, USA, 2001.
15. Goodby, J.W. Symmetry and Chirality in Liquid Crystals. In *Handbook of Liquid Crystals*, 1st ed.; Demus, D., Goodby, J.W., Gray, G.W., Spiess, H.-W., Vill, V., Eds.; Wiley-VCH: Weinheim, Germany, 1998; Volume 1, pp. 115–132.
16. Seki, A.; Funahashi, M. Photovoltaic Effects in Ferroelectric Liquid Crystals based on Phenylterthiophene Derivatives. *Chem. Lett.* **2016**, *45*, 616–618. [CrossRef]

17. Seki, A.; Funatsu, Y.; Funahashi, M. Anomalous photovoltaic effect based on molecular chirality: Influence of enantiomeric purity on the photocurrent response in π-conjugated ferroelectric liquid crystals. *Phys. Chem. Chem. Phys.* **2017**, *19*, 16446–16455. [CrossRef] [PubMed]
18. Seki, A.; Funatsu, Y.; Funahashi, M. Chiral photovoltaic effect in an ordered smectic phase of a phenylterthiophene derivative. *Org. Electron.* **2018**, *62*, 311–319. [CrossRef]
19. Seki, A.; Yoshio, M.; Mori, Y.; Funahashi, M. Ferroelectric Liquid-Crystalline Binary Mixtures Based on Achiral and Chiral Trifluoromethylphenylterthiophenes. *ACS Appl. Mater. Interfaces* **2020**, *12*, 53029–53038. [CrossRef]
20. Mulder, D.J.; Schenning, A.P.H.J.; Bastiaansen, C.W.M. Chiral-nematic liquid crystals as one dimensional photonic materials in optical sensors. *J. Mater. Chem. C* **2014**, *2*, 6695–6705. [CrossRef]
21. Hartmann, W.J.A.M. Ferroelectric Liquid Crystal Displays for Television Application. *Ferroelectrics* **1991**, *122*, 1–26. [CrossRef]
22. Funahashi, M. Chiral Liquid Crystalline Electronic Systems. *Symmetry* **2021**, *13*, 672. [CrossRef]
23. Shimizu, Y.; Oikawa, K.; Nakayama, K.; Guillon, D. Mesophase semiconductors in field effect transistors. *J. Mater. Chem.* **2007**, *17*, 4223–4229. [CrossRef]
24. Pisula, W.; Zorn, M.; Chang, J.Y.; Müllen, K.; Zentel, R. Liquid Crystalline Ordering and Charge Transport in Semiconducting Materials. *Macromol. Rapid Commun.* **2009**, *30*, 1179–1202. [CrossRef] [PubMed]
25. Funahashi, M. Development of Liquid-Crystalline Semiconductors with High Carrier Mobilities and Their Application to Thin-film Transistors. *Polym. J.* **2009**, *41*, 459–469. [CrossRef]
26. O'Neill, M.; Kelly, S.M. Ordered Materials for Organic Electronics and Photonics. *Adv. Mater.* **2011**, *23*, 566–584. [CrossRef] [PubMed]
27. Seki, A.; Funahashi, M. Nanostrucure Formation Based on the Functionalized Side Chains in Liquid-Crystalline Heteroaromatic Compounds. *Heterocycles* **2016**, *92*, 3–30.
28. Kato, T.; Yoshio, M.; Ichikawa, T.; Soberats, B.; Ohno, H.; Funahashi, M. Transport of ions and electrons in nanostructured liquid crystals. *Nat. Rev. Mater.* **2017**, *2*, 17001. [CrossRef]
29. Hori, T.; Miyake, Y.; Yamasaki, N.; Yoshida, H.; Fujii, A.; Shimizu, Y.; Ozaki, M. Solution Processable Organic Solar Cell Based on Bulk Heterojunction Utilizing Phthalocyanine Derivative. *Appl. Phys. Express* **2010**, *3*, 101602. [CrossRef]
30. Shin, W.; Yasuda, T.; Watanabe, G.; Yang, Y.S.; Adachi, C. Self-Organizing Mesomorphic Diketopyrrolopyrrole Derivatives for Efficient Solution-Processed Organic Solar Cells. *Chem. Mater.* **2013**, *25*, 2549–2556. [CrossRef]
31. Hassheider, T.; Benning, S.A.; Kitzerow, H.-S.; Achard, M.-F.; Bock, H. Color-Tuned Electroluminescence from Columnar Liquid Crystalline Alkyl Arenecarboxylates. *Angew. Chem. Int. Ed.* **2001**, *40*, 2060–2063. [CrossRef]
32. Aldred, M.P.; Contoret, A.E.A.; Farrar, S.R.; Kelly, S.M.; Mathieson, D.; O'Neill, M.; Tsoi, W.C.; Vlachos, P. A Full-Color Electroluminescent Device and Patterned Photoalignment Using Light-Emitting Liquid Crystals. *Adv. Mater.* **2005**, *17*, 1368–1372. [CrossRef]
33. Benning, S.A.; Oesterhaus, R.; Kitzerow, H.-S. Polarized electroluminescence of a discotic mesogenic compound. *Liq. Cryst.* **2004**, *31*, 201–205. [CrossRef]
34. van Breemen, A.J.J.M.; Herwig, P.T.; Chlon, C.H.T.; Sweelssen, J.; Schoo, H.F.M.; Setayesh, S.; Hardeman, W.M.; Martin, C.A.; de Leeuw, D.M.; Valeton, J.J.P.; et al. Large Area Liquid Crystal Monodomain Field-Effect Transistors. *J. Am. Chem. Soc.* **2006**, *128*, 2336–2345. [CrossRef] [PubMed]
35. Funahashi, M.; Zhang, F.; Tamaoki, N. High Ambipolar Mobility in a Highly Ordered Smectic Phase of a Dialkylphenylterthiophene Derivative That Can Be Applied to Solution-Processed Organic Field-Effect Transistors. *Adv. Mater.* **2007**, *19*, 353–358. [CrossRef]
36. Pisula, W.; Menon, A.; Stepputat, M.; Lieberwirth, I.; Kolb, U.; Tracz, A.; Sirringhaus, H.; Pakula, T.; Müllen, K. A Zone-Casting Technique for Device Fabrication of Field-Effect Transistors Based on Discotic Hexa-*peri*-hexabenzocoronene. *Adv. Mater.* **2005**, *17*, 684–689. [CrossRef]
37. Funatsu, Y.; Sonoda, A.; Funahashi, M. Ferroelectric liquid-crystalline semiconductors based on a phenylterthiophene skeleton: Effect of introduction of oligosiloxane moieties and photovoltaic effect. *J. Mater. Chem. C* **2015**, *3*, 1982–1993. [CrossRef]
38. Glass, A.M.; von der Linde, D.; Negran, T.J. High-voltage bulk photovoltaic effect and the photorefractive process in $LiNbO_3$. *Appl. Phys. Lett.* **1974**, *25*, 233–235. [CrossRef]
39. Fridkin, V.M.; Popov, B.N. The Anomalous Photovoltaic Effect and Photoconductivity in Ferroelectrics. *Phys. Status Solidi A* **1978**, *46*, 729–733. [CrossRef]
40. Yuan, Y.; Xiao, Z.; Yang, B.; Huang, J. Arising applications of ferroelectric materials in photovoltaic devices. *J. Mater. Chem. A* **2014**, *2*, 6027–6041. [CrossRef]
41. Butler, K.T.; Frost, J.M.; Walsh, A. Ferroelectric materials for solar energy conversion: Photoferroics revisited. *Energy Environ. Sci.* **2015**, *8*, 838–848. [CrossRef]
42. Sotome, M.; Nakamura, M.; Fujioka, J.; Ogino, M.; Kaneko, Y.; Morimoto, T.; Zhang, Y.; Kawasaki, M.; Nagaosa, N.; Tokura, Y.; et al. Spectral dynamics of shift current in ferroelectric semiconductor SbSI. *Proc. Natl. Acad. Sci. USA* **2019**, *116*, 1929–1933. [CrossRef]
43. Morimoto, T.; Nakamura, M.; Kawasaki, M.; Nagaosa, N. Current-Voltage Characteristic and Shot Noise of Shift Current Photovoltaics. *Phys. Rev. Lett.* **2018**, *121*, 267401. [CrossRef] [PubMed]

44. Nakamura, M.; Hatada, H.; Kaneko, Y.; Ogawa, N.; Tokura, Y.; Kawasaki, M. Impact of electrodes on the extraction of shift current from a ferroelectric semiconductor SbSI. *Appl. Phys. Lett.* **2018**, *113*, 232901. [CrossRef]
45. Sugita, A.; Suzuki, K.; Tasaka, S. Ferroelectric properties of a triphenylene derivative with polar functional groups in the crystalline state. *Phys. Rev. B* **2004**, *69*, 212201. [CrossRef]
46. Sasabe, H.; Nakayama, T.; Kumazawa, K.; Miyata, S.; Fukuda, E. Photovoltaic Effect in Poly(vinylidene fluoride). *Polym. J.* **1981**, *13*, 967–973. [CrossRef]
47. Zhang, C.; Nakano, K.; Nakamura, M.; Araoka, F.; Tajima, K.; Miyajima, D. Noncentrosymmetric Columnar Liquid Crystals with the Bulk Photovoltaic Effect for Organic Photodetectors. *J. Am. Chem. Soc.* **2020**, *142*, 3326–3330. [CrossRef]
48. Funahashi, M. High open-circuit voltage under the bulk photovoltaic effect for the chiral smectic crystal phase of a double chiral ferroelectric liquid crystal doped with a fullerene derivative. *Mater. Chem. Front.* **2021**, *5*, 8265–8274. [CrossRef]
49. Wurzbach, I.; Rothe, C.; Bruchlos, K.; Ludwigs, S.; Giesselmann, F. Shear alignment and 2D charge transport of tilted smectic liquid crystalline phases—XRD and FET studies. *J. Mater. Chem. C* **2019**, *7*, 2615–2624. [CrossRef]
50. Yelamaggad, C.V.; Shashikala, I.S.; Hiremath, U.S.; Shankar Rao, D.S.; Prasad, S.K. Liquid crystal dimers possessing chiral rod-like anisometric segments: Synthesis, characterization and electro-optic behaviour. *Liq. Cryst.* **2007**, *34*, 153–167. [CrossRef]
51. Sund, P.; Pettersson, F.; Österbacka, R.; Wilén, C.-E. Conductivity, interaction and solubility of hetero-bifunctional end-capped β,β′-dihexylsubstituted sexithiophenes. *J. Appl. Polym. Sci.* **2018**, *135*, 46830. [CrossRef]
52. Swamy, K.C.K.; Kumar, N.N.B.; Balaraman, E.; Kumar, K.V.P.P. Mitsunobu and Related Reactions: Advances and Applications. *Chem. Rev.* **2009**, *109*, 2551–2651. [PubMed]
53. Shi, Y.-J.; Hughes, D.L.; McNamara, J.M. Stereospecific synthesis of chiral tertiary alkyl-aryl ethers via Mitsunobu reaction with complete inversion of configuration. *Tetrahedron Lett.* **2003**, *44*, 3609–3611. [CrossRef]
54. Bouchta, A.; Nguyen, H.T.; Achard, M.F.; Hardouin, F.; Destrade, C.; Twieg, R.J.; Maaroufi, A.; Isaert, N. New TGB$_A$ series exhibiting a $S_C{}^*S_AS_A{}^*N^*$ phase sequence. *Liq. Cryst.* **1992**, *12*, 575–591. [CrossRef]
55. Hall, A.W.; Lacey, D.; Hill, J.S.; Blackwood, K.M.; Jones, M.; McDonnell, D.G.; Sage, I.C. Synthesis and evaluation of a series of novel 2-substituted poly(allylalcohol) side chain liquid crystalline oligomers exhibiting ferroelectricity. *Liq. Cryst.* **1996**, *20*, 437–447. [CrossRef]
56. McCubbin, A.J.; Snieckus, V.; Lemieux, R.P. Ferroelectric liquid crystals with fluoro- and aza-fluorenone cores: The effect of stereo-polar coupling. *Liq. Cryst.* **2005**, *32*, 1195–1203. [CrossRef]
57. Sackmann, H.; Demus, D. The Problems of Polymorphism in Liquid Crystals. *Mol. Cryst. Liq. Cryst.* **1973**, *21*, 239–273. [CrossRef]
58. Funahashi, M.; Hanna, J. High ambipolar carrier mobility in self-organizing terthiophene derivative. *Appl. Phys. Lett.* **2000**, *76*, 2574–2576. [CrossRef]
59. Rao, P.B.; Rao, N.V.S.; Pisipati, V.G.K.M. The Smectic F Phase in nO.m Compounds. *Mol. Cryst. Liq. Cryst.* **1991**, *206*, 9–15. [CrossRef]
60. Ouchi, Y.; Uemura, T.; Takezoe, H.; Fukuda, A. Molecular Reorientaion Process in Chiral Smectic I Liquid Crystal. *Jpn. J. Appl. Phys.* **1985**, *24*, 893–895. [CrossRef]
61. Uemoto, T.; Yoshino, K.; Inuishi, Y. Electrical and Optical Properties of Ferroelectric Liquid Crystals and Influence of Applied Pressure. *Mol. Cryst. Liq. Cryst.* **1981**, *67*, 137–152. [CrossRef]
62. Doi, T.; Sakurai, Y.; Tamatani, A.; Takenaka, S.; Kusabayashi, S.; Nishihata, Y.; Terauchi, H. Thermal and X-Ray Diffraction Studies of Liquid Crystals incorporating a Perfluoroalkyl Group. *J. Mater. Chem.* **1991**, *1*, 169–173. [CrossRef]
63. Lee, M.; Cho, B.-K.; Kim, H.; Yoon, J.-Y.; Zin, W.-C. Self-Organization of Rod–Coil Molecules with Layered Crystalline States into Thermotropic Liquid Crystalline Assemblies. *J. Am. Chem. Soc.* **1998**, *120*, 9168–9179. [CrossRef]
64. Ting, C.-H.; Hsu, C.-S. Synthesis and photoluminescence property of polyacetylenes containing liquid crystalline side groups. *J. Polym. Res.* **2001**, *8*, 159–167. [CrossRef]
65. Ting, C.-H.; Chen, J.-T.; Hsu, C.-S. Synthesis and Thermal and Photoluminescence Properties of Liquid Crystalline Polyacetylenes Containing 4-Alkanyloxyphenyl trans-4-Alkylcyclohexanoate Side Groups. *Macromolecules* **2002**, *35*, 1180–1189. [CrossRef]
66. Vergara-Toloza, R.O.; Soto-Bustamante, E.A.; Gonzalez-Henriquez, C.M.; Haase, W. Side chain liquid crystalline composites, occurrence of interdigitated bilayer smectic C phases. *Liq. Cryst.* **2011**, *38*, 911–916. [CrossRef]
67. Zhu, P.; Wu, L.; Liu, W.; Li, B.; Li, Y.; Yang, Y. Methylation driven molecular packing difference at the smectic phases of a series of pyridinium-based chiral ionic liquid crystals. *Mol. Cryst. Liq. Cryst.* **2021**, *731*, 55–65. [CrossRef]
68. Hsu, H.-F.; Chien, S.-J.; Chen, H.-H.; Chen, C.-H.; Huang, L.-Y.; Kuo, C.-H.; Chen, K.-J.; Ong, C.W.; Wong, K.-T. Mesogenic and optical properties of α,α′-bis(4-alkoxyphenylethynyl)oligothiophenes. *Liq. Cryst.* **2005**, *32*, 683–689. [CrossRef]

Chromonic Ionic Liquid Crystals Forming Nematic and Hexagonal Columnar Phases

Takahiro Ichikawa *, Mei Kuwana and Kaori Suda

Department of Biotechnology, Tokyo University of Agriculture and Technology, Nakacho, Koganei, Tokyo 184-8588, Japan
* Correspondence: t-ichi@cc.tuat.ac.jp

Abstract: We designed an ionic salt by combining a π-conjugated anion and a cholinium cation. It formed homogeneous mixtures with water in various weight ratios. The obtained mixtures showed chromonic liquid-crystalline behavior in a wider concentration range as compared to analogous compounds with inorganic cations. Although only an exhibition of nematic phases was previously reported by Kasianova et al. for analogous compounds with an inorganic cation in 2010, the ionic salt with a cholinium cation showed not only nematic phases but also hexagonal columnar phases. The formation of hexagonal columnar phases is attributed to its ability to form mesophases even in a high concentration range, which enables the cylindrical aggregates of the π-conjugated anions to form dense packing. By examining the states of the water molecules, we revealed that the ability of the cholinium cation to form a hydrated ionic liquid state strongly contributes to the widening of the concentration range forming chromonic liquid-crystalline behavior.

Keywords: chromonic liquid crystal; ionic liquid; nematic; hexagonal columnar phase

1. Introduction

Chromonic liquid crystals are a class of lyotropic liquid crystals. A unique point of chromonic liquid crystals is that they have molecular structures composed of a polycyclic aromatic core having several ionic and/or hydrophilic groups at its periphery [1–8]. It has been generally understood that the aromatic core plays a key role for the formation of self-assembled cylindrical aggregates through π-π interactions and/or other interactions. The hydrophilic groups are important for solubility into water. To date, there have been many reports of ionic compounds exhibiting chromonic liquid-crystalline (LC) behavior. Most of them are composed of π-conjugated mesogens with anionic groups and inorganic cations [1–6] while, in some case, chromonic liquid crystals composed of π-conjugated cations and inorganic anions have been also reported [7,8].

On the other hand, in the several decades of studies, there have been a growing interest on the use of organic cations for creating functional ionic compounds, such as ionic liquids [9–11], ionic liquid crystals [12–15], ionic plastic crystals [16–18], and ionic crystals [19]. In the course of studies on ionic liquids, it has been revealed that there is a potential that a slight difference of the organic cation structures results in the large difference of physicochemical properties and functions. For example, imidazolium cations are recognized as one of the most suitable cations for designing ionic liquids dissolving cellulose [20,21]. On the other hands, the use of cholinium cation has attracted an increasing attention for yielding hydrated ionic liquids for dissolving some bio-functional polymers [22]. For example, Fujita and Ohno reported that hydrated ionic liquids have a great potential as liquid media for enzyme storage [23–25]. One of the important characteristics of hydrated ionic liquids is that they maintain liquid states even in quite high concentration conditions. This characteristic leads us to envision that the employment of suitable organic cations would be one of an advanced strategy for controlling the chromonic LC behavior of π-conjugated compounds with anionic groups.

As an anion with π-conjugated structure, we have employed 4,4-(5,5-dioxidodibenzo[b,d]thiene-3,7-diyl)dibenzenesulphonic acid (**pQpdS**) anion. It is an anion whose Cs salt was reported to exhibit chromonic LC behavior at a water content of 85 wt% by Kasianova et al. in 2010 [26]. As an organic cation, a cholinium (Ch) cation has been selected. By combining these cation and anion, we have synthesized an organic salt, **pQpdS-Ch** (Figure 1). Its chromonic LC behavior in water has been examined using polarized optical microscopy (POM) observation, differential scanning calorimetry (DSC), and X-ray diffraction (XRD) measurements.

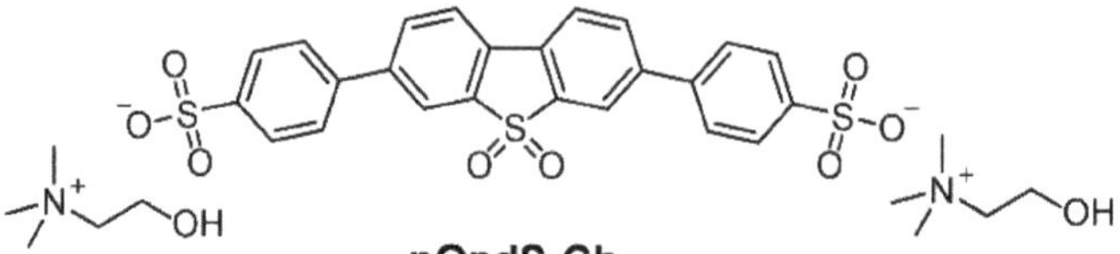

Figure 1. Molecular structure of **pQpdS-Ch**.

2. Materials and Methods

The synthesis scheme of **pQpdS-H** is shown in Scheme 1. To an aqueous solution of choline hydroxide, an equimolar amount of 4,4-(5,5-dioxidodibenzo[b,d]thiene-3,7-diyl)dibenzenesulphonic acid (**pQpdS-H**) was added. The solution was stirred until the white solid of **pQpdS-H** dissolved into the solution. Evaporation of water yielded a **pQpdS-Ch** as a white solid.

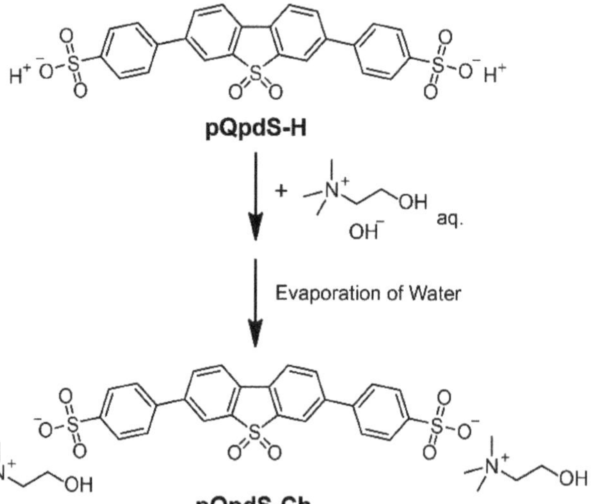

Scheme 1. Synthesis of **pQpdS-Ch**.

^{1}H NMR (400MHz, D$_2$O): δ = 7.78 (s, 2H), 7.64 (d, J = 8.4 Hz, 4H), 7.41–7.35 (m, 8H), 3.92–3.89 (m, 4H), 3.37–3.35 (m, 4H), 3.05 (s, 18H).

3. Results and Discussion

Mixtures of **pQpdS-Ch** and H$_2$O in 100–X:X weight ratios (X = 90, 80, 70, 60, and 50) were prepared by adding two components into Eppendorf tubes. In order to increase the homogeneity of the mixtures, the tubes were vibrated and centrifugation was performed.

We could obtain the homogeneous mixtures when 90 ≥ X ≥ 50 while it was not obtained when X ≤ 40. Small amounts of the homogeneous mixtures were put on a slide glass and covered with a cover glass. Polarizing optical microscope (POM) observation was carried out for them while cooling from isotropic phases observed at around 80 °C. The obtained textures are shown in Figure 2. It has been found that the samples with X ≥ 90 shows no birefringence in the temperature range, indicating that mesomorphic behavior is not induced when X ≥ 90. On the other hand, the mixture with X = 80 shows a schlieren texture, which is a characteristic of nematic phases. This behavior is similar to that reported by Kasianova et al. for **pQpdS** with Cs cation [26]. A notable difference has been observed when X ≤ 70. These mixtures show focal conic fan-textures, which are indicative of the formation of columnar LC phases. The thermotropic phase transition behavior of the mixtures is summarized in a bar graph (Figure 3). The formation of the nematic phases results from the cylinder aggregation of the **pQpdS** anions and the subsequent axial alignment of the cylinders. That of the columnar phases can be attributed to the formation of the positional order of the cylinders as the decrease of the inter-cylinder distance.

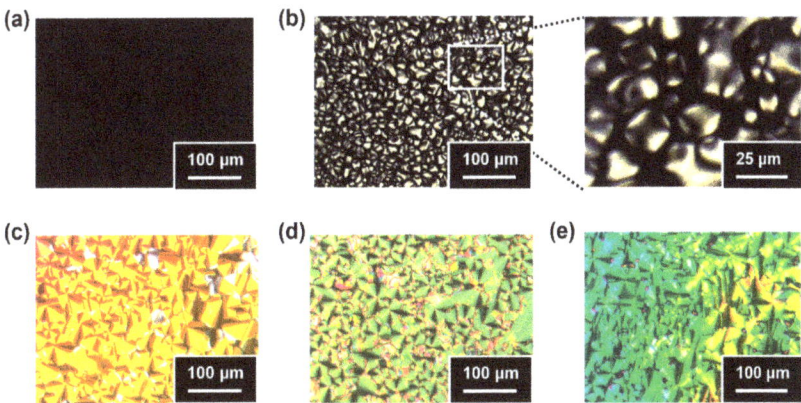

Figure 2. Polarized optical microscopic images of the mixtures of **pQpdS-Ch** and H_2O in the 100–X:X weight ratios. (**a**) X = 90 at 25 °C, (**b**) X = 80 at 10 °C, (**c**) X = 70 at 25 °C, (**d**) X = 60 at 25 °C, and (**e**) X = 50 at 25 °C.

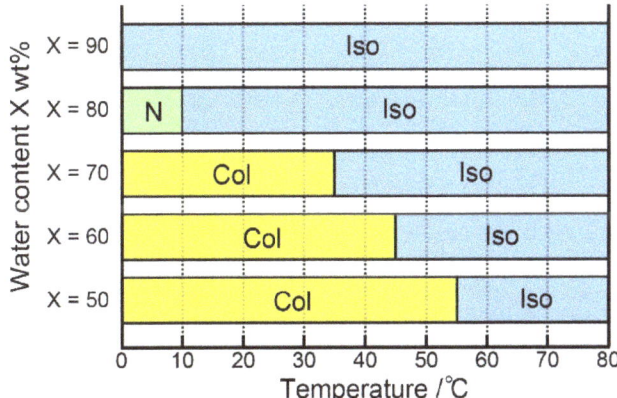

Figure 3. Bar graph of the thermotropic phase transition behavior of the mixtures of **pQpdS-Ch** and H_2O in the 100–X:X weight ratios on cooling.

In order to confirm the formation of columnar phases, we have performed XRD measurements for the mixtures at 30 °C. For avoiding the evaporation of water from the sample, the LC samples were put on an aluminium pan and rapidly covered by a polymer film (DURA SEAL, DIVERSIFIED BIOTECH). A XRD pattern observed for the mixture (X = 60) is shown in Figure 4. An intense peak and two weak peaks were found in the small angle region. The d-values estimated from the peak position θ values are 32.7, 18.7, and 16.0 Å, respectively. These peaks can be indexed as (100), (110), and (200) reflections of a hexagonal structure, which lead us to identify the columnar phase as a hexagonal columnar (Col$_h$) phase. The intercolumnar distances in the Col$_h$ LC phase can be calculated to be 37.8 Å from the d-values. The formation of Col$_h$ phases for bent shaped chromonic liquid crystals has been also reported by Wang et al. in 2018 [27], which supports our characterization.

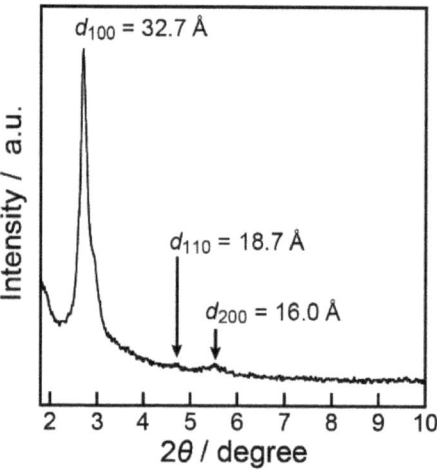

Figure 4. X-ray diffraction (XRD) pattern of **pQpdS-Ch**/H$_2$O (X = 60) weight ratio at 30 °C.

POM observation for macroscopically aligned samples is a useful strategy for deducing the molecular assembled structures in LC states. In order to employ this strategy for the present materials, we sandwiched a small amount of a **pQpdS-Ch**/H$_2$O (X = 60) mixture between a cover glass and a slide glass and then added a mechanical shearing to the cover glass. It is a technique to align the column axis to the shearing direction [28]. As expected, the sheared sample show a homogeneous texture under POM observation, which is indicative of the formation of 1D-aligned columnar phases. The aligned samples were observed under POM with a 530 nm retardation plate inserted in the optical path at 45 degrees. The shearing direction is set parallel and perpendicular to the slow axis direction of the retardation plate. It has been found that, when these two directions are parallel, the texture is observed in a yellow (Figure 5a). It turns into in a blue as the rotation of the sample through 90 degrees (Figure 5b). These results mean that the slow axis of the Col$_h$ liquid crystals is perpendicular to the column axis, namely the **pQpdS-Ch**/H$_2$O (X = 60) mixture has a negative birefringence. It is consistent with the results obtained for analogous compound with the Cs cation in a nematic phase that was reported by Kasianova et al. [26].

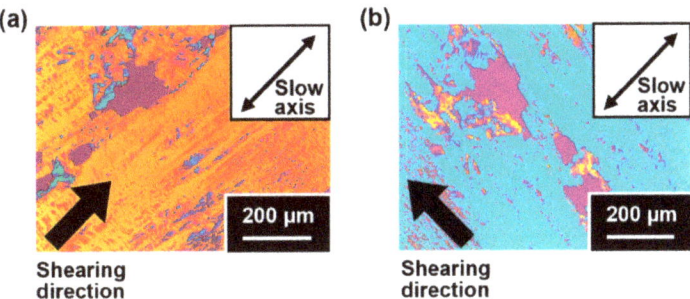

Figure 5. Polarized optical microscopic images of the mixtures of **pQpdS-Ch**/H_2O (X = 60) in the Col$_h$ phase after shearing. (**a**) The shearing direction is parallel to the slow axis of the retardation plate. (**b**) The shearing direction is perpendicular to the slow axis of the retardation plate.

In order to further confirm the molecular assembled structure of the **pQpdS-Ch**/H_2O mixtures in the Col$_h$ phase, we have performed polarized IR measurements. A 1D-aligned sample of the **pQpdS-Ch**/H_2O (X = 70) mixture was prepared by the same method. IR absorbance was measured with setting the angle of the polarizer (θ_p) in the range from 0 to 180 degrees. The Col$_h$ LC sample was set in such a way that its column axis was parallel to θ_p = 0 degree. While the S=O stretching vibration ($\nu_{S=O}$) of the **pQpdS** molecules was observed at 1301 cm^{-1} independent of the θ_p angles, the peak strength of $\nu_{S=O}$ clearly depended on the θ_p angles. For example, the absorbance of $\nu_{S=O}$ was 0.20 when θ_p = 0 degree, which increases as the increase of θ_p (Figure 6a). For further clarify θ_p-dependence of the $\nu_{S=O}$ absorbance, we have constructed a polar plot (Figure 6b). It indicates that the sulfonyl groups of the **pQpdS** molecules are oriented perpendicular to the 1D column axis.

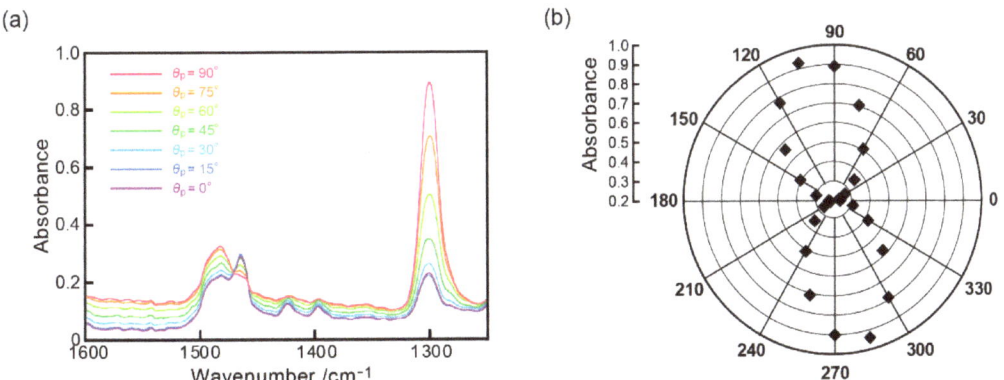

Figure 6. (**a**) θ_p-dependence of IR spectra of the **pQpdS-Ch**/H_2O (X = 70) mixture in the Col$_h$ phase. (**b**) A polar plot of the absorbance of the S=O stretching vibration ($\nu_{S=O}$) observed at 1301 cm^{-1}.

The phase transition behavior of these mixtures has been further examined using DSC measurements. The DSC measurements were performed in the temperature range from 0 to 80 °C at the heating/cooling rate of 10 °C min^{-1}. The obtained DSC charts on cooling and heating are shown in Figure 7a,b, respectively. In the cooling process, an exothermic peak is found for each sample when X ≤ 80. These peaks can be attributed to the enthalpy change at the phase transition from an isotropic phase to an LC phase. It can be seen that the peak position shifts to higher temperature region as the decrease of X, which is consistent with the isotropization temperatures observed by POM observation. The thermal stabilization of the mesophases upon the decrease of X can be explained by the increase of the packing density of the cylinder aggregates. Another notable trend is that the peak area increases as the decrease of X. For example, the peak area of the phase transition from the Col_h to Iso phases is 0.83 mJ/mg for the **pQpdS-Ch**/H_2O (X = 70) mixture while it increases to 2.16 mJ/mg for **pQpdS-Ch**/H_2O (X = 50). These enthalpy changes can be mainly ascribed to the cleavage of the dipole–dipole interaction between the sulfonyl groups. Rough calculation is described in the Supplemental Information.

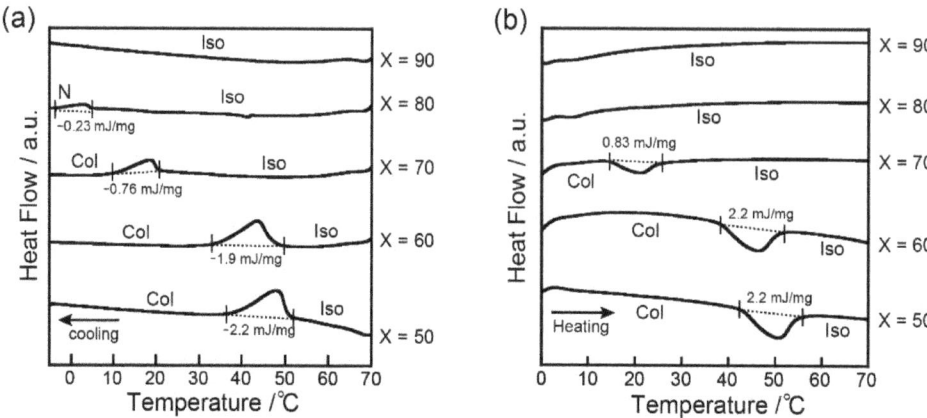

Figure 7. DSC thermograms of the **pQpdS-Ch**/H_2O mixtures: (**a**) on cooling and (**b**) on heating.

It is considered that the water molecules in the mixtures exist as bound water and/or free water. With an aim to investigate the states of water in the mixtures, we have performed DSC measurements in lower temperature region. The DSC measurements were carried out from room temperature to −80 °C. A peak corresponding to crystallization of free water was found at a temperature lower than 0 °C. By estimating the amount of free water in the mixtures from the peak area, the molar ratios of bound water and free water in the mixtures (X = 50–90) were investigated and summarized in Table 1. It has been found that 15–25 water molecules strongly interact with a **pQpdS-Ch** molecule and then exist as bound water. These results are consistent with the number of hydration water molecules reported for cholinium-based hydrated ionic liquids [29]. We assume that the cylindrical aggregates formed by the **pQpdS** anions are surrounded by sheath of hydrated ionic liquids that produce liquidity and prevent crystallization even in the water poor condition (70 ≥ X ≥ 50).

Table 1. Weight ratios and molar ratios of **pQpdS-Ch**, free water, and bound water.

Water Content X	Exothermic Peak Area (mJ/mg)	Component Ratios in Weight			Component Ratios in Mole		
		pQpdS-Ch	Free Water	Bound Water	pQpdS-Ch	Free Water	Bound Water
90	256	10	77	13	1	313	55
80	228	20	38	12	1	139	24
70	184	30	55	15	1	75	20
60	135	40	40	20	1	41	20
50	106	50	32	18	1	26	15

Based on the results of POM observation, DSC, and XRD measurements, here we discuss the molecular assembled structure of the **pQpdS-Ch**/H_2O mixtures. For assuming the molecular assembled structures, an important characteristic of **pQpdS-Ch** is that it has a strong dipole moment at the sulfonyl group, which can be calculated to be 5.2 D by DFT calculation (Figure S1) (see supplementary materials). Therefore, it is expected that it forms a dimer in the dissolved state as well as in the assembled states in water. The size of the **pQpdS-Ch** anion is about 20 Å. Based on these results, here we imagine a molecular assembled structure of the **pQpdS-Ch**/H_2O (X = 60) mixture in the Col$_h$ phase. The number of bound water per the dimer of the **pQpdS-Ch** molecules calculated from the endothermic peaks in the DSC charts is 20×2 = 40. The inter columnar distance is calculated to be 37.8 Å as explained in the above paragraph. Considering these data and the component weight ratio, the molecular assembled structures of the **pQpdS-Ch**/H_2O (X = 60) mixture in the Col$_h$ phase is drawn as shown in Figure 8.

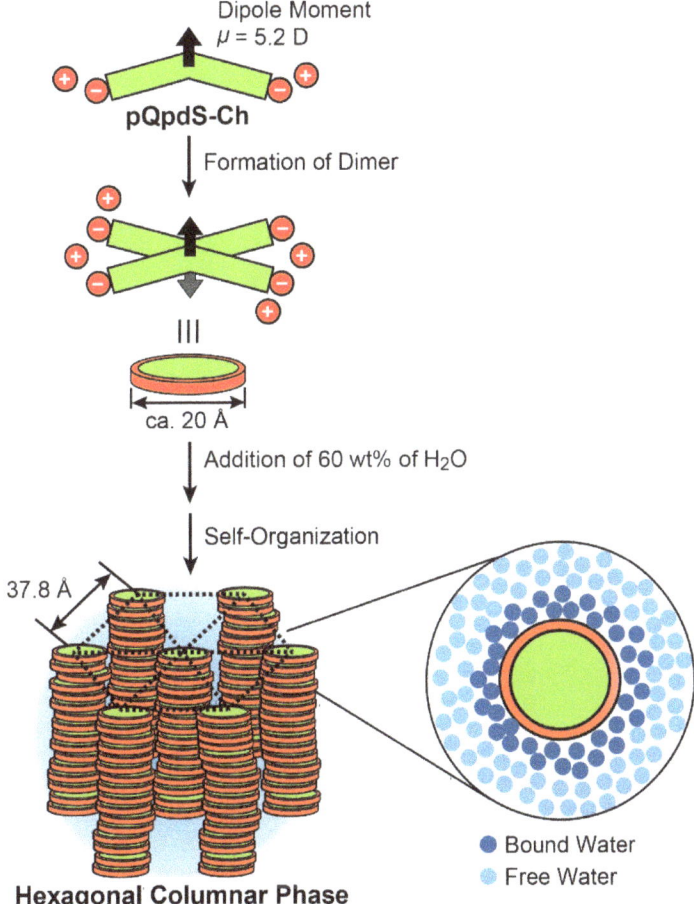

Figure 8. A schematic image of the molecular assembled structure of the **pQpdS-Ch**/H_2O (X = 60) mixture in the Col$_h$ phase.

In order to further confirm the effects of the cation species, we have also prepared analogous compounds with other inorganic cations, such as Li, Na, and K cations. **pQpdS-Y** (Y = Li, Na, and K) were prepared according to the same procedure used for **pQpdS-Ch**. They were obtained as white or slightly yellowish white compounds (Figure S2). The mixtures of these compounds and water were prepared with varying the component ratios and their phase transition behavior was examined by POM observation.

It has been found that the exhibition of N phases was observed for the **pQpdS-Li**/H_2O mixtures when the water content value is $90 \geq X \geq 85$ and that of Col$_h$ phases was observed when X = 80 (Figure 9). The water content dependence of the mesophase pattern is similar to that of the **pQpdS-Ch**/H_2O mixtures. These results indicate that the formation of Col$_h$ is a phenomena that is observed not solely for **pQpdS-X** with organic cations but also for **pQpdS-X** with inorganic.

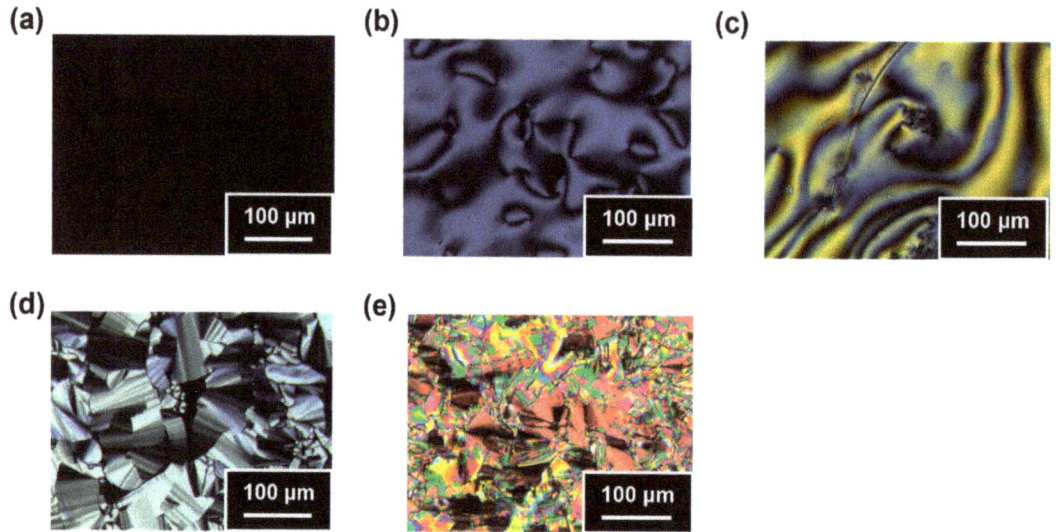

Figure 9. Polarized optical microscopic images of the **pQpdS-Li**/H_2O mixtures in the 100–X:X weight ratios. (**a**) X = 95 at 25 °C, (**b**) X = 90 at 20 °C, (**c**) X = 85 at 25 °C, (**d**) X = 80 at 50 °C, and (**e**) X = 70 at 25 °C.

On the other hand, we have found that the **pQpdS-Na**/H_2O mixtures forms only N phases when $95 \geq X \geq 80$ (Figure 10) and those with $70 \geq X$ form crystalline states. Comparing the water content range forming mesophases for the **pQpdS-X**/H_2O mixtures (Figures 3 and 11), it can be seen that the employment of the cholinium cation provides chromonic liquid crystals showing LC behavior in the widest water content range. It is attributed to the higher solubility of **pQpdS-Ch** into water that results from its lower crystallinity than those with inorganic cations. Namely, the employment of the cholinium cation increases the conformational degrees of freedom, which contributes to the inhibition of the crystallization. The melting point of **pQpdS-Ch** is higher than 100 °C (Figure S3) that is the important temperature of the definition of ionic liquids. However, considering the recent studies on ionic liquids where hydrated organic salts are called hydrated ionic liquids [23–25,30], we expect that **pQpdS-Ch**/H_2O mixture can be regarded as hydrated ionic liquids exhibiting chromonic LC behavior.

In the course of studies on ionic liquids, they have been used in a wide range of fields, including electrochemistry, analysis, catalysis, and solvents. Focusing on hydrated ionic liquids, they have been expected as potential solvent for biomolecules [23,31]. On the other hand, chromonic liquid crystals have been investigated as sensors [32] and optical materials [33]. We believe that the present material design will attract attention in a wide field of research ranging from biotechnology to material chemistry.

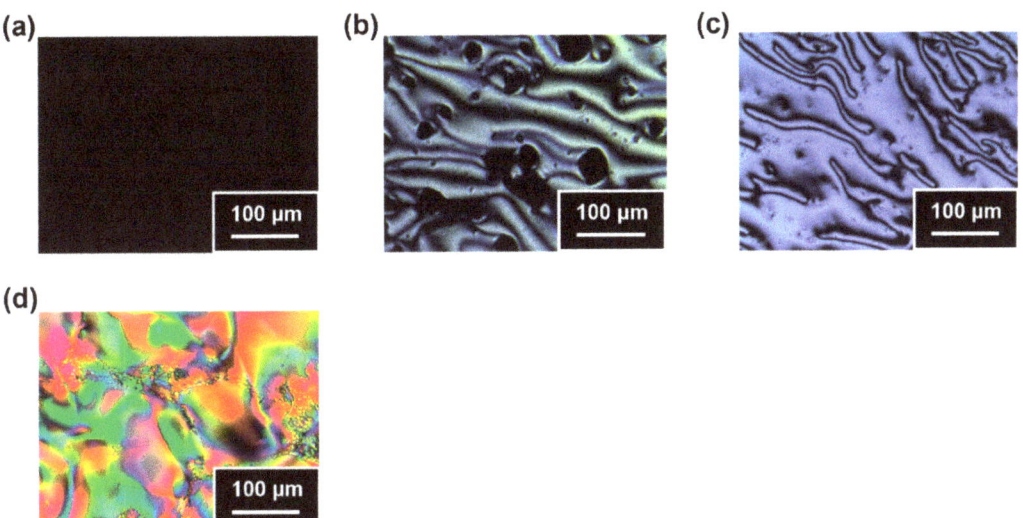

Figure 10. Polarized optical microscopic images of the **pQpdS-Na**/H_2O mixtures in the 100–X:X weight ratios. (**a**) X = 95 at 25 °C, (**b**) X = 90 at 25 °C, (**c**) X = 85 at 25 °C, and (**d**) X = 80 at 25 °C.

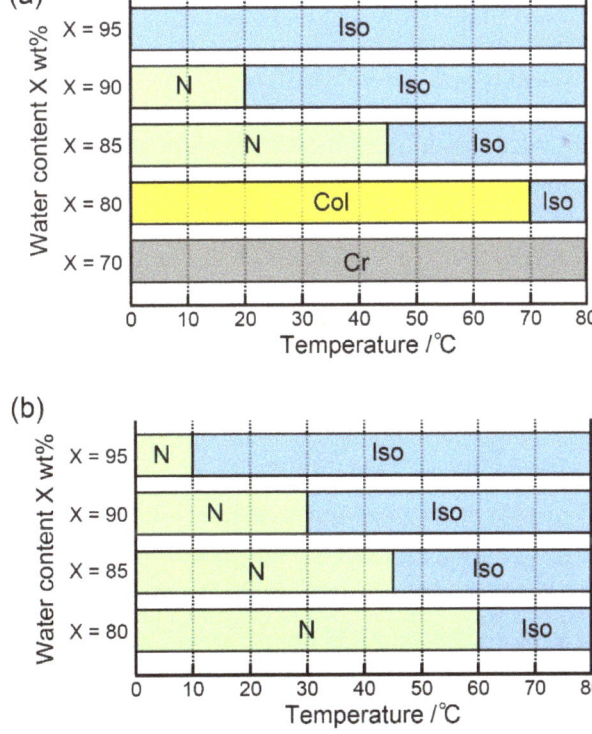

Figure 11. Bar graphs of the thermotropic phase transition behavior of; (**a**) the **pQpdS-Li**/H_2O mixtures on cooling; (**b**) the **pQpdS-Na**/H_2O mixtures on cooling.

4. Conclusions

We have succeeded in preparing a new class of an organic salt **pQpdS-Ch** showing chromonic liquid-crystalline (LC) behavior. This compound is composed of a rod-shaped aromatic anion with a strong dipole moment and cholinium cations. Both of two components owes their specific tasks. The former plays a key role for the formation of cylindrical aggregates via dipole–dipole interactions. The latter plays an important role for the formation of hydrated states. Moreover, since the hydrated cholinium cation has a larger positional and conformational degrees of freedom than inorganic cations, it results in the inhibition of the crystallization of the cylindrical aggregates. This effect enables to form chromonic LC mesophases even in a wider concentration range than a series of analogous compounds with inorganic cations, which leads to the exhibition of chromonic hexagonal columnar phases.

Supplementary Materials: The following supporting information can be downloaded at: https://www.mdpi.com/article/10.3390/cryst12111548/s1, Figure S1: A schematic image of the molecular assembled structure. Figure S2: Pictures of the synthesized compounds; Figure S3: TG/DTA result.

Author Contributions: T.I., M.K. and K.S. conceived and designed the experiments; T.I., M.K. and K.S. performed the experiments and analyzed the data; T.I., M.K. and K.S. wrote the paper. All authors have read and agreed to the published version of the manuscript.

Funding: This research was funded by JSPS KAKENHI numbers JP21H02010, and JP22H04526 from the Japan Society for the Promotion of Science.

Data Availability Statement: Not applicable.

Acknowledgments: This work was supported by JSPS KAKENHI numbers JP21H02010, and JP22H04526 from the Japan Society for the Promotion of Science. This work was partly supported by the financial support from FUJIFILM.

Conflicts of Interest: The authors declare no conflict of interest.

References

1. Tam-Chang, S.-W.; Huang, L. Chromonic liquid crystals: Properties and applications as functional materials. *Chem. Commun.* **2008**, *17*, 1957–1967. [CrossRef]
2. Collings, P.J.; Dickinson, A.J.; Smith, E.C. Molecular aggregation and chromonic liquid crystals. *Liq. Cryst.* **2010**, *37*, 701–710. [CrossRef]
3. Lydon, J. Chromonic liquid crystalline phases. *Liq. Cryst.* **2011**, *38*, 1663–1681. [CrossRef]
4. Zhou, S. Recent progresses in lyotropic chromonic liquid crystal research: Elasticity, viscosity, defect structures, and living liquid crystals. *Liq. Cryst. Today* **2018**, *27*, 91–108. [CrossRef]
5. Ruslim, C.; Matsunaga, D.; Hashimoto, M.; Tamaki, T.; Ichimura, K. Structural Characteristics of the Chromonic Mesophases of C.I. Direct Blue 67. *Langmuir* **2003**, *19*, 3686–3691. [CrossRef]
6. Hara, M.; Nagano, S.; Mizoshita, N.; Seki, T. Chromonic/Silica Nanohybrids: Synthesis and Macroscopic Alignment. *Langmuir* **2007**, *23*, 12350–12355. [CrossRef]
7. Iverson, I.K.; Tam-Chang, S.-W. Cascade of Molecular Order by Sequential Self-Organization, Induced Orientation, and Order Transfer Processes. *J. Am. Chem. Soc.* **1999**, *121*, 5801–5802. [CrossRef]
8. Tam-Chang, S.-W.; Iverson, I.K.; Helbley, J. Study of the Chromonic Liquid-Crystalline Phases of Bis-(N,N-diethylaminoethyl) perylene-3,4,9,10-tetracarboxylic Diimide Dihydrochloride by Polarized Optical Microscopy and 2H NMR Spectroscopy. *Langmuir* **2004**, *20*, 342–347. [CrossRef]
9. Ohno, H. Functional Design of Ionic liquids. *Bull. Chem. Soc. Jpn.* **2006**, *79*, 1665–1680. [CrossRef]
10. Ohno, H.; Yoshizawa-Fujita, M.; Kohno, Y. Functional Design of Ionic Liquids Unprecedented Liquids that Contribute to Energy Technology, Bioscience, and Materials Sciences. *Bull. Chem. Soc. Jpn.* **2019**, *92*, 852–868. [CrossRef]
11. MacFarlane, D.R.; Forsyth, M.; Howlett, P.C.; Kar, M.; Passerini, S.; Pringle, J.M.; Ohno, H.; Watanabe, M.; Yan, F.; Zheng, W.; et al. Ionic liquids and their solid-state analogues as materials for energy generation and storage. *Nat. Rev. Mater.* **2016**, *1*, 15005. [CrossRef]
12. Kato, T.; Yoshio, M.; Ichikawa, T.; Soberats, B.; Ohno, H.; Funahashi, M. Transport of ions and electrons in nanostructured liquid crystals. *Nat. Rev. Mater.* **2017**, *2*, 17001. [CrossRef]
13. Goossens, K.; Lava, K.; Bielawski, C.W.; Binnemans, K. Ionic Liquid Crystals: Versatile Materials. *Chem. Rev.* **2016**, *116*, 4643–4807. [CrossRef]
14. Axenov, K.V.; Laschat, S. Thermotropic Ionic Liquid Crystals. *Materials* **2011**, *4*, 206–259. [CrossRef]

15. Ichikawa, T.; Kato, T.; Ohno, H. Dimension control of ionic liquids. *Chem. Commun.* **2019**, *55*, 8205–8214. [CrossRef]
16. Pringle, J.M.; Howlett, P.C.; MacFarlane, D.R.; Forsyth, M. Organic ionic plastic crystals: Recent advances. *J. Mater. Chem.* **2010**, *20*, 2056–2062. [CrossRef]
17. MacFarlane, D.R.; Huang, J.; Forsyth, M. Lithium-doped plastic crystal electrolytes exhibiting fast ion conduction for secondary batteries. *Nature* **1999**, *402*, 792–794. [CrossRef]
18. Yamaguchi, S.; Yamada, H.; Takeoka, Y.; Rikukawa, M.; Yoshizawa-Fujita, M. Synthesis of pyrrolidinium-based plastic crystals exhibiting high ionic conductivity at ambient temperature. *New J. Chem.* **2019**, *43*, 4008–4012. [CrossRef]
19. Shimoyama, Y.; Uchida, S. Structure-function Relationships of Porous Ionic Crystals (PICs) Based on Polyoxometalate Anions and Oxo-centered Trinuclear Metal Carboxylates as Counter Cations. *Chem. Lett.* **2021**, *50*, 21–30. [CrossRef]
20. Swatloski, R.P.; Spear, S.K.; Holbrey, J.D.; Rogers, R.D. Dissolution of Cellose with Ionic Liquids. *J. Am. Chem. Soc.* **2002**, *124*, 4974–4975. [CrossRef]
21. Fukaya, Y.; Hayashi, K.; Wada, M.; Ohno, H. Cellulose dissolution with polar ionic liquids under mild conditions: Required factors for anions. *Green Chem.* **2008**, *10*, 44–46. [CrossRef]
22. Fujita, K.; Ohno, H. Stable G-quadruplex structure in a hydrated ion pair: Cholinium cation and dihydrogen phosphate anion. *Chem. Commun.* **2012**, *48*, 5751–5753. [CrossRef]
23. Fujita, K.; MacFarlane, D.R.; Forsyth, M.; Yoshizawa-Fujita, M.; Murata, K.; Nakamura, N.; Ohno, H. Solubility and stability of cytochrome c in hydrated ionic liquids: Effect of oxo acid residues and kosmotropicity. *Biomacromolecules* **2007**, *8*, 2080–2086. [CrossRef]
24. Fujita, K.; Sanada, M.; Ohno, H. Sugar chain-binding specificity and native folding state of lectins preserved in hydrated ionic liquids. *Chem. Commun.* **2015**, *51*, 10883–10886. [CrossRef]
25. Fujita, K.; Nakano, R.; Nakaba, R.; Nakamura, N.; Ohno, H. Hydrated ionic liquids enable both solubilisation and refolding of aggregated concanavalin A. *Chem. Commun.* **2019**, *55*, 3578–3581. [CrossRef]
26. Kasianova, I.; Kharatyian, E.; Geivandov, A.; Palto, S. Lyotropic liquid crystal guest-host material and anisotropic thin films for optical applications. *Liq. Cryst.* **2010**, *37*, 1439–1451. [CrossRef]
27. Wang, D.; Yan, Q.; Zhong, F.; Li, Y.; Fu, M.; Meng, L.; Huang, Y.; Li, L. Counterion-Induced Nanosheet-to-Nanofilament Transition of Lyotropic Bent-Core Liquid Crystals. *Langmuir* **2018**, *34*, 13006–13013. [CrossRef]
28. Yoshio, M.; Mukai, T.; Ohno, H.; Kato, T. One-Dimensional Ion Transport in Self-Organized Columnar Ionic Liquids. *J. Am. Chem. Soc.* **2004**, *126*, 994–995. [CrossRef]
29. Ohno, H.; Fujita, K.; Kohno, Y. Is seven a minimum number of water molecules per ion pair for assured biological activity in ionic liquid/water mixtures? *Phys. Chem. Chem. Phys.* **2015**, *17*, 14454–14460. [CrossRef]
30. Haberler, M.; Schröder, C.; Steinhauser, O. Hydrated Ionic Liquids with and without Solute: The Influence of Water Content and Protein Solutes. *J. Chem. Theory Comput.* **2012**, *8*, 3911–3928. [CrossRef]
31. Fujita, K.; MacFarlane, D.R.; Forsyth, M. Protein solubilising and stabilising ionic liquids. *Chem. Commun.* **2005**, *38*, 4804–4806. [CrossRef]
32. Shiyanovskii, S.V.; Lavrentovich, O.D.; Schneider, T.; Ishikawa, T.; Smalyukh, I.I.; Woolverton, C.J.; Niehaus, G.D.; Doane, K.J. Lyotropic Chromonic Liquid Crystals for Biological Sensing Applications. *Mol. Cryst. Liq. Cryst.* **2005**, *434*, 259–270. [CrossRef]
33. Bosire, R.; Ndaya, D.; Kasi, R.M. Recent progress in functional materials from lyotropic chromonic liquid crystals. *Polym. Int.* **2021**, *70*, 938–943. [CrossRef]

Article

Twist–Bend Nematic Phase Behavior of Cyanobiphenyl-Based Dimers with Propane, Ethoxy, and Ethylthio Spacers

Yuki Arakawa *, Yuto Arai, Kyohei Horita, Kenta Komatsu and Hideto Tsuji

Department of Applied Chemistry and Life Science, Graduate School of Engineering, Toyohashi University of Technology, 1-1 Hibarigaoka, Tempaku-cho, Toyohashi 441-8580, Japan
* Correspondence: arakawa@tut.jp

Abstract: The twist–bend nematic (N_{TB}) phase is a liquid crystal (LC) phase with a heliconical structure that typically forms below the temperature of the conventional nematic (N) phase. By contrast, the direct transition between the N_{TB} and isotropic (Iso) phases without the intermediation of the N phase rarely occurs. Herein, we demonstrate the effects of linkage type (i.e., methylene, ether, and thioether) on the typical Iso–N–N_{TB} and rare direct Iso–N_{TB} phase-transition behaviors of cyanobiphenyl (CB) dimers CB3CB, CB2OCB, and CB2SCB bearing three-atom-based propane, ethoxy, and ethylthio spacers, respectively. In our previous study, CB2SCB exhibited the monotropic direct Iso–N_{TB} phase transition. In this study, we report that CB3CB also shows the direct Iso–N_{TB} phase transition, whereas CB2OCB exhibits the typical Iso–N–N_{TB} phase sequence with decreasing temperature. The Iso–LC (Iso–N_{TB} or Iso–N) phase-transition temperatures upon cooling show the order CB2OCB (108 °C) > CB3CB (49 °C) > CB2SCB (43 °C). The thioether-linked CB2SCB is vitrifiable, whereas CB3CB and CB2OCB exhibit strong crystallization tendencies. The phase-transition behaviors are also discussed in terms of the three bent homologous series with different oligomethylene spacers n: CBnCB, CBnOCB, and CBnSCB.

Keywords: twist–bend nematic phase; direct twist–bend nematic phase transition; liquid crystal dimer; cyanobiphenyl dimer; short spacer

1. Introduction

Liquid crystal (LC) phases are mesophases between anisotropic crystals and isotropic liquids. A nematic (N) phase generally does not have an apparent layered structure and is recognized as the most fluid LC phase. After the heliconical twist–bend nematic (N_{TB}) phase was predicted [1–3], it was experimentally assigned to unknown N (N_X) phases below the conventional N phase of bent molecules (dimers) in the last decade [4,5]. The N_{TB} phase possesses a heliconical director precession with a pitch of approximately 10 nm [6–8]. The heliconical structures of the N_{TB} phase result in optical textures and physical properties like those of layered smectic (Sm) phases rather than the conventional N phase [9–13]. Therefore, the N_{TB} phase is often considered a pseudo-layered phase. However, the phase identification of the N_{TB} phase for the N_X phase is still under discussion and further study is required to elucidate the detailed structure owing to its elusive nature. Alternatively, a polar twisted nematic (N_{PT}) phase model for the N_X phase instead of the N_{TB} phase has been proposed [14,15]. In this paper, the widely recognized term, N_{TB} phase, is used.

The N_{TB} phase can be formed only by bent molecules, such as bent dimers [4,5,16–39], linear oligomers (e.g., trimers, tetramers, and hexamers) [40–49], duplexed hexamers [50], polymers [51], hydrogen-bonded dimers and trimers [52,53] with odd-number atom spacers, and bent-core molecules [54–56]. Mandle reviewed the structure–property relationship of the N_{TB} phase and summarized the recent progress in this topic [57]. Theoretical simulation studies have examined the relationship between the curvature of various bent dimers and the incidence of the N_{TB} phase [58–61]. In nearly all cases of the reported bent

molecules, the N_{TB} phase continuously formed at a temperature below the temperature of the conventional N phase, resulting in a typical isotropic (Iso)–N–N_{TB} phase sequence with decreasing temperature. For example, the homologous series of bent symmetric methylene- and asymmetric methylene-/ether- and methylene-/thioether-linked cyanobiphenyl (CB) dimers with n number of carbon atoms in the oligomethylene spacers, CBnCB (n = 5, 7, 9, 11, and 13) [17,25], CBnOCB (n = 4, 6, 8, and 10) [24,25], and CBnSCB (n = 4, 6, 8, and 10) [62], respectively, are known to exhibit the typical Iso–N–N_{TB} phase sequence (Figure 1a).

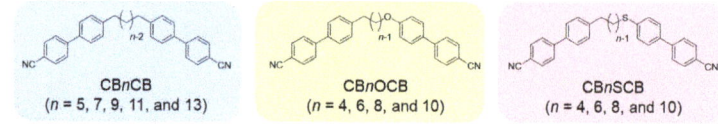

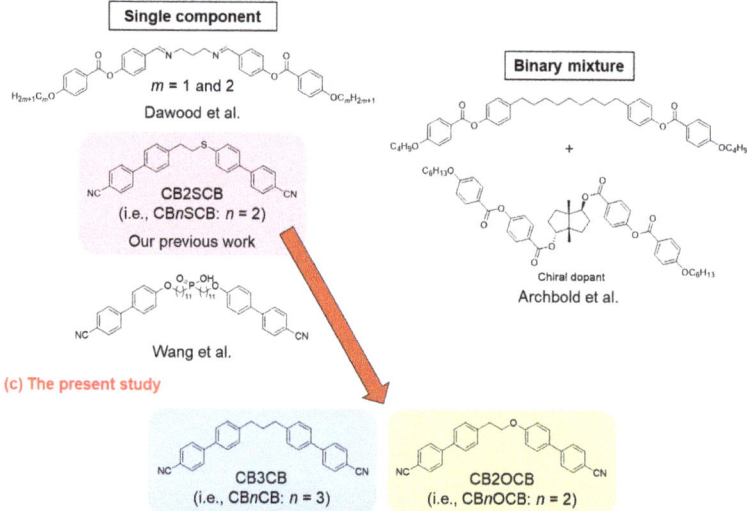

Figure 1. Bent dimer systems exhibiting the typical Iso–N–N_{TB} and rare direct Iso–N_{TB} phase transitions investigated in previous studies and the present study. (**a**) Previous work on bent-shaped CB dimer homologs with longer spacers, including methylene-linked CBnCB (odd n = 5, 7, 9, 11, and 13) [17,25], methylene-/ether-linked CBnOCB (even n = 4, 6, 8, and 10) [24,25], and methylene-/thioether-linked CBnSCB (n = 4, 6, 8, and 10) [62] that exhibit the typical Iso–N–N_{TB} phase sequence. (**b**) Previous work on dimers showing the rare Iso–N_{TB} phase sequence from binary mixtures reported by Archbold et al. [63] (**right**) and single-component dimers reported by Dawood et al. [64,65] (imine-linked dimers, **top left**), our group [62] (ethylthio-linked CB2SCB, **center left**), and Wang et al. [66] (a phosphine-bridged dimer, **bottom left**). (**c**) Single-component dimers CB3CB and CB2OCB in the present study.

However, in a few cases, an N_{TB} phase can be directly formed from the Iso phase without the intermediate N phase. Archbold et al. reported that binary mixtures of a dimer with the Iso–N–N_{TB} phase sequence and a chiral dopant (6–10 wt%) exhibit the direct Iso–N_{TB} phase transition (Figure 1b, right) [63]. Dawood et al. reported that two imine-linked dimers with a central propane spacer and terminal methoxy or ethoxy groups (m = 1 and 2, respectively) (Figure 1b, top left) show the direct Iso–N_{TB} phase transition [64,65]. We reported that bent CBnSCB (n = 4, 6, 8, and 10) demonstrates the Iso–N–N_{TB} phase sequence, as described earlier, whereas only the shortest ethylthio-linked CB2SCB exhibits the direct Iso–N_{TB} phase transition upon cooling (Figure 1b, center left) [62]. Shortening the flexible spacer n of the CBnSCB dimers lowers the Iso–N phase-transition temperature (T_{IN}) upon cooling, thereby narrowing the intermediate N-phase temperature range (ΔT_N), which leads to the direct Iso–N$_{TB}$ phase transition of CB2SCB. Moreover, Wang et al. recently disclosed that a phosphorus-bridged LC dimer exhibits the direct Iso–N$_{TB}$ phase transition (Figure 1b, bottom left) [66].

Considering that, in our previous study, CB2SCB demonstrated the rare direct Iso–N$_{TB}$ phase transition [62], propane-linked CB3CB and ethoxy-linked CB2OCB bearing the same three-atom-based spacers (as shown in Figure 1c) are worthy of further investigation whether they exhibit the rare Iso–N$_{TB}$ phase or not. The influence of the different linkage types on the phase transition of such short-spacer dimers has yet to be reported. In this study, we investigated the phase-transition behaviors of three CB-based dimers with different three-atom-based spacers, namely CB3CB, CB2OCB, and CB2SCB. The phase-transition behavior of the previously reported CB3CB that did not show the LC phase [67] was reinvestigated, while that of CB2OCB was explored for the first time. The phase-transition data of CB2SCB were obtained from our previous study [62]. We then compared the phase-transition behaviors of three series of CBnCB, CBnOCB, and CBnSCB homologs. Finally, the effects of linkage-type on the N$_{TB}$ phase-transition behavior of dimers with short spacers, particularly on the occurrence of the direct Iso–N$_{TB}$ phase transition, were explored.

2. Materials and Methods

2.1. General

The synthetic routes of CB3CB and CB2OCB are shown in Scheme 1. The molecular structures were analyzed using ^{1}H and ^{13}C nuclear magnetic resonance (NMR) spectroscopy on a JNM ECX 500 spectrometer (JEOL Ltd., Tokyo, Japan). Phase identification was conducted via polarized optical microscopy (POM) using a BX50 microscope (Olympus Corp., Tokyo, Japan) on an LK-600 PM hot stage (Linkam, Surrey, UK). The phase-transition temperatures and associated enthalpy changes were determined using differential scanning calorimetry (DSC) on a DSC-60 Plus (Shimadzu Corp., Kyoto, Japan). Calibration was performed using indium, and the measurements were performed over heating/cooling/heating cycles at a rate of 10 °C min^{-1} under an N$_2$ gas flow (50 mL min^{-1}).

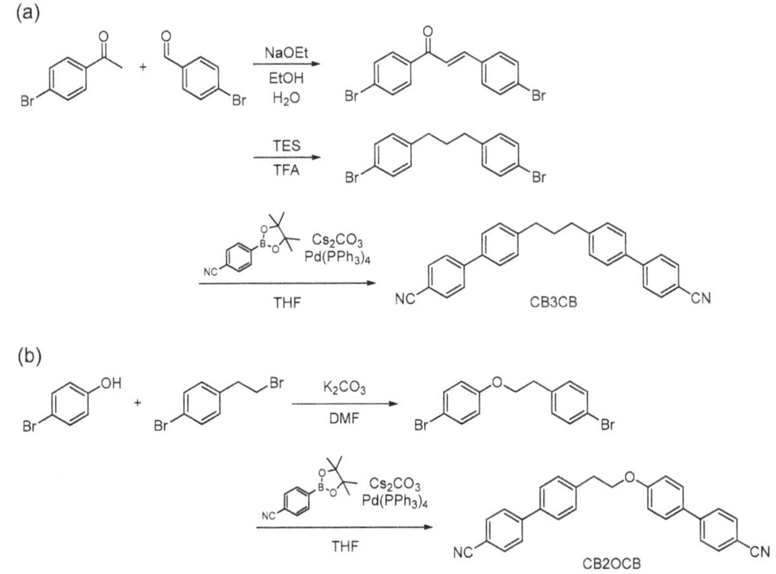

Scheme 1. Synthesis of (**a**) CB3CB and (**b**) CB2OCB.

2.2. Synthesis

2.2.1. 1,3-Bis(4-cyanobiphenyl-4′-yl)propane (CB3CB)

(*E*)-1,3-Bis(4-bromophenyl)prop-2-en-1-one

This compound was synthesized referring to a previously reported method [68]. 4′-Bromoacetophenone (2.19 g, 11.0 mmol), *p*-bromobenzaldehyde (2.04 g, 11.0 mmol), sodium ethoxide (NaOEt) (1.12 g, 16.5 mmol), ethanol (EtOH) (30 mL), and distilled water (10 mL) were added to a round-bottom flask. The resultant mixture was stirred at ambient temperature for 1 h. The reaction mixture was filtered, and the residue was rinsed with copious amounts of methanol to afford the target compound as a pale-yellow solid (86%). ^{1}H NMR (500 MHz, CDCl$_3$) δ 7.89 (d, *J* = 9.0 Hz, Ar–*H*, 2H), 7.75 (d, *J* = 16.0 Hz, CO–C*H*, 1H), 7.65 (d, *J* = 9.0 Hz, Ar–*H*, 2H), 7.56 (d, *J* = 9.0 Hz, Ar–*H*, 2H), 7.51 (d, *J* = 9.0 Hz, Ar–*H*, 2H), 7.47 (d, *J* = 16.0 Hz, CO–CH=C*H*, 1H) ppm. ^{13}C NMR (126 MHz, CDCl$_3$) δ 189.1, 143.9, 136.7, 133.7, 132.3, 132.0, 130.0, 129.8, 128.1, 125.1, 121.9 ppm.

1,3-Bis(4-bromophenyl)propane

This compound was also synthesized referring to the literature [68]. (*E*)-1,3-Bis(4-bromophenyl)prop-2-en-1-one (1.10 g, 2.99 mmol) was added to a two-necked round-bottom flask, which was then purged with argon gas. Trifluoroacetic acid (TFA) (11 mL) was then added into the flask, followed by adding triethylsilane (TES) (4.8 mL, 30 mmol) dropwise. The resultant mixture was stirred at ambient temperature for 1 h. TES (0.95 mL, 5.94 mmol) was added to the reaction mixture, which was further stirred for 18 h. The reaction mixture was poured into distilled water, extracted with dichloromethane, and washed with brine. The obtained solution was then dried over magnesium sulfate (MgSO$_4$), and the solvent was removed under reduced pressure. The residue was purified by column chromatography on silica gel using hexane as an eluent to afford 1,3-bis(4-bromophenyl)propane as a colorless solid (78%). ^{1}H NMR (500 MHz, CDCl$_3$) δ 7.39 (d, *J* = 8.5 Hz, Ar–*H*, 4H), 7.04 (d, *J* = 8.5 Hz, Ar–*H*, 4H), 2.58 (t, *J* = 7.5 Hz, Ar–C*H*$_2$, 4H), 1.90 (tt, *J* = 7.5 and 7.5 Hz, Ar–CH$_2$–C*H*$_2$, 2H) ppm. ^{13}C NMR (126 MHz, CDCl$_3$) δ 140.9, 131.4, 130.2, 119.5, 34.6, 32.6 ppm.

CB3CB

1,3-Bis(4-bromophenyl)propane (200 mg, 0.565 mmol), 4-(4,4,5,5-tetramethyl-1,3,2-dioxaborolan-2-yl)benzonitrile (267 mg, 1.17 mmol), cesium carbonate (Cs$_2$CO$_3$) (748 mg, 2.29 mmol), and tetrakis(triphenylphosphine)palladium(0) [Pd(PPh$_3$)$_4$] (41.5 mg, 35.9 µmol) were added to a two-necked round-bottom flask, which was then purged with argon gas. Tetrahydrofuran (THF) (5 mL) was degassed by bubbling with argon gas and added to the flask. The resultant mixture was stirred at reflux temperature for 3 h. Subsequently, Pd(PPh$_3$)$_4$ (39.5 mg, 34.2 µmol) was added to the reaction mixture, which was further stirred for 4 h. The reaction mixture was extracted with dichloromethane and washed with brine. The solution was then dried over MgSO$_4$, and the volatile solvent was removed under reduced pressure. The residue was purified by column chromatography on silica gel using dichloromethane/hexane (1:1, *v*/*v*) and recrystallized in a dichloromethane/hexane mixture to afford CB3CB as a colorless solid (51%). ^{1}H NMR (500 MHz, CDCl$_3$) δ 7.71 (d, *J* = 8.5 Hz, Ar–*H*, 4H), 7.67 (d, *J* = 8.5 Hz, Ar–*H*, 4H), 7.53 (d, *J* = 8.5 Hz, Ar–*H*, 4H), 7.31 (d, *J* = 8.5 Hz, Ar–*H*, 4H), 2.74 (t, *J* = 7.5 Hz, Ar–C*H*$_2$, 4H), 2.04 (tt, *J* = 7.5 and 7.5 Hz, Ar–CH$_2$–C*H*$_2$, 2H) ppm. ^{13}C NMR (126 MHz, CDCl$_3$) δ 145.5, 142.9, 136.7, 132.6, 129.2, 127.5, 127.2, 119.0, 110.6, 35.0, 32.7 ppm.

2.2.2. (4-Cyanobiphenyl-4′-yloxy)-2-(4-cyanobiphenyl-4′-yl)ethane (CB2OCB)

1-Bromo-4-[2-(4-bromophenoxy)ethyl]benzene

4-Bromophenethyl bromide (690 mg, 2.61 mmol), 4-bromophenol (302 mg, 1.75 mmol), and potassium carbonate (K$_2$CO$_3$) (617 mg, 4.46 mmol) were added to a two-necked round-bottom flask, which was then purged with argon gas. *N*,*N*-Dimethylformamide (DMF) (5 mL) was degassed by bubbling argon gas and then added to the flask. The resultant mixture was stirred at 90 °C for 18 h. More 4-bromophenethyl bromide (690 mg, 2.61 mmol) was added to the reaction mixture, which was further stirred for 5 h. The reaction mixture was extracted with ethyl acetate and washed with brine. The obtained solution was dried over MgSO$_4$, and then the solvent was removed under reduced pressure. The residue was purified by column chromatography on silica gel using hexane/ethyl acetate (20:1, v/v) to afford 1-Bromo-4-[2-(4-bromophenoxy)ethyl]benzene (28%). ^{1}H NMR (500 MHz, CDCl$_3$) δ 7.43 (d, J = 8.5 Hz, Ar–H, 2H), 7.35 (d, J = 8.5 Hz, Ar–H, 2H), 7.14 (d, J = 8.5 Hz, Ar–H, 2H), 6.75 (d, J = 8.5 Hz, Ar–H, 2H), 4.11 (t, J = 6.5 Hz, Ar–O–CH_2, 2H), 3.03 (t, J = 6.8 Hz, Ar–O–CH$_2$–CH_2, 2H) ppm. ^{13}C NMR (126 MHz, CDCl$_3$) δ 157.8, 137.1, 132.3, 131.6, 130.7, 120.4, 116.3, 113.0, 68.5, 35.1 ppm.

CB2OCB

1-Bromo-4-[2-(4-bromophenoxy)ethyl]benzene (96.7 mg, 0.272 mmol), 4-(4,4,5,5-tetra methyl-1,3,2-dioxaborolan-2-yl)benzonitrile (140 mg, 0.611 mmol), Cs$_2$CO$_3$ (182 mg, 0.559 mmol), and Pd(PPh$_3$)$_4$ (64.1 mg, 55.5 µmol) were added to a two-necked round-bottom flask, which was then purged with argon gas. THF (3 mL) was degassed by bubbling argon gas and then added to the flask. The resultant mixture was stirred at reflux temperature for 16 h. An arbitrary amount of Pd(PPh$_3$)$_4$ was added to the reaction mixture. The mixture was then stirred for 16 h, extracted with dichloromethane, and washed with brine. The solution was dried over MgSO$_4$, and the volatile solvent was removed under reduced pressure. The residue was purified by column chromatography on silica gel using dichloromethane/hexane (5:1, v/v) and recrystallized in a dichloromethane/hexane mixture to afford CB2OCB (20%). ^{1}H NMR (500 MHz, CDCl$_3$) δ 7.72 (d, J = 8.5 Hz, Ar–H, 2H), 7.69 (d, J = 8.5 Hz, Ar–H, 2H), 7.67(d, J = 8.0 Hz, Ar–H, 2H), 7.63 (d, J = 8.5 Hz, Ar–H, 2H), 7.55 (d, J = 8.5 Hz, Ar–H, 2H), 7.52 (d, J = 9.0 Hz, Ar–H, 2H), 7.42 (d, J = 8.0 Hz, Ar–H, 2H), 7.00 (d, J = 9.0 Hz, Ar–H, 2H), 4.27 (t, J = 6.5 Hz, Ar–O–CH_2, 2H), 3.18 (t, J = 6.5 Hz, Ar–O–CH$_2$–CH_2, 2H) ppm. ^{13}C NMR (126 MHz, CDCl$_3$) δ 159.3, 145.3, 145.1, 138.9, 137.5, 132.59, 132.56, 131.7, 129.8, 128.4, 127.5, 127.3, 127.1, 119.1, 118.9, 115.1, 110.8, 110.1, 68.5, 35.3 ppm.

3. Results

3.1. Phase-Transition Behaviors of CB3CB, CB2OCB, and CB2SCB

As shown in the DSC curves (Figure 2a), the methylene-linked CB3CB sample exhibited a melting temperature (T_m) of ~149 °C, where it transitioned to the Iso phase without forming the LC phase upon heating. Upon cooling, the CB3CB sample crystallized at ~80 °C from the Iso phase. These phase-transition temperatures were higher than those (142.1 and 69.1 °C, respectively) reported in the literature [67], where CB3CB did not exhibit the LC phase. However, POM observations in this study revealed that in the supercooled Iso phase of CB3CB (Figure 3a), which does not undergo crystallization at ~80 °C, birefringent textures appear (Figure 3b), which then grow mixed textures including fan-, focal-conic-, and rope-like domains, as shown in Figure 3c,d. Besides, we did not observe typical N-phase textures such as marble and schlieren textures during this phase transition, as seen in Figure 3. This texture behavior is similar to the direct Iso–N$_{TB}$ phase transition of CB2SCB [62]. Therefore, it was revealed that CB3CB also shows the monotropic direct Iso–N$_{TB}$ phase transition at ~49 °C. During the heating after the cooling, the observed N$_{TB}$ phase of CB3CB transitioned to the Iso phase at ~55 °C. The CB3CB sample displayed a strong crystallization tendency and did not vitrify upon cooling, even at a higher rate of 30 °C min^{-1}. Additionally, the ether-linked CB2OCB did not show LC phases over the first and second heating cycles, where it exhibited different T_m values of 164.4 and 139.7 °C, respectively; Figure 2b represents the latter. This result indicates the existence of crystal polymorphs that depend on the crystallization conditions. Upon cooling, most of the Iso-phase domains and droplets of CB2OCB crystallize at ~104 °C, as shown by the exothermic peak in Figure 2b. However, the POM images reveal the formation of the N and N$_{TB}$ phases at ~108 and 78 °C, respectively, in the supercooled Iso phase, as confirmed by the marble/schlieren textures (Figure 4a) and blocky texture (Figure 4b), respectively. Because of the strong crystallization tendencies of CB3CB and CB2OCB, as shown by the DSC curves (Figure 2a,b, respectively), their LC phases were not investigated by X-ray diffractometry. The strong crystallization tendency of these molecules differs from the vitrifiable CB2SCB, with a glass transition temperature of ~20 °C, as shown in Figure 2c [62]. The T_m values upon the second heating, the associated enthalpy changes (ΔH), and the Iso–N$_{TB}$, Iso–N, and N–N$_{TB}$ phase-transition temperatures upon the cooling (T_{INTB}, T_{IN}, and T_{NNTB}, respectively) of CB3CB, CB2OCB, and CB2SCB are summarized in Table 1. For simplicity, the crystallization and glass transition temperatures upon cooling are not listed in Table 1.

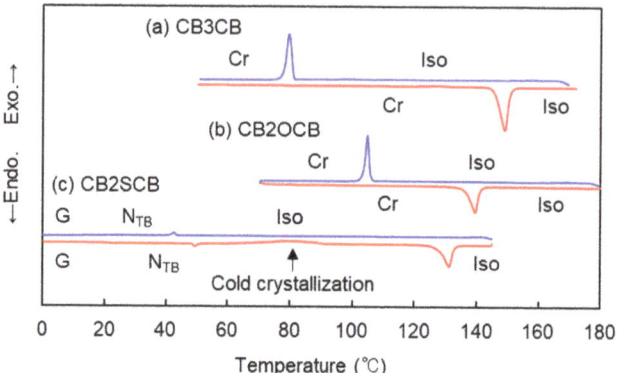

Figure 2. DSC curves of (**a**) CB3CB, (**b**) CB2OCB, and (**c**) CB2SCB upon the second heating (red lines) and cooling (blue lines) cycles at a rate of 10 °C min^{-1}. Cr and G denote the crystal phase and glassy state, respectively. Panel (**c**) is reproduced from Ref. [62].

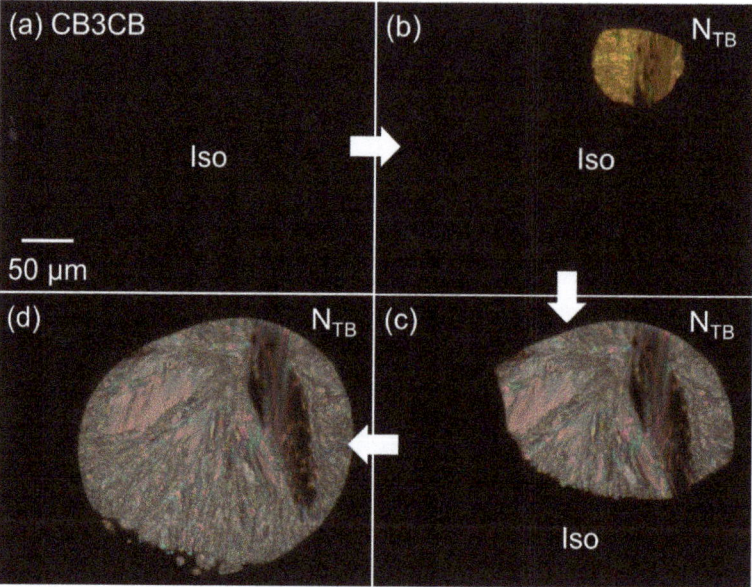

Figure 3. POM images during the Iso–N$_{TB}$ phase transition of CB3CB: (**a**) the Iso phase (52 °C), (**b**) the N$_{TB}$ texture appearance in the Iso phase (50 °C), (**c,d**) growth in the N$_{TB}$ texture (48.5 and 47.5 °C, respectively).

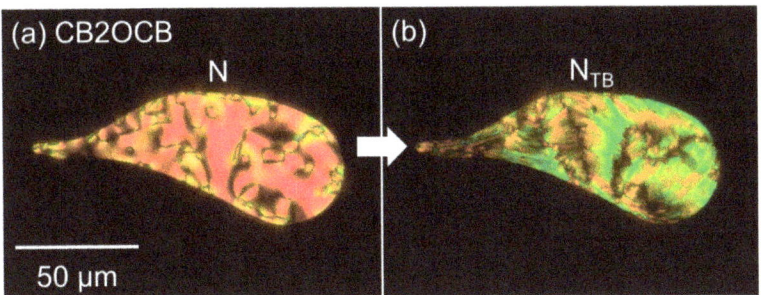

Figure 4. POM images of (**a**) the N phase (104 °C) and (**b**) the N_{TB} phase (69 °C) of CB2OCB.

Table 1. T_m and the associated ΔH upon second heating and T_{INTB}, T_{IN}, and T_{NNTB} upon cooling for CB3CB, CB2OCB, and CB2SCB.

Sample	Heating	Cooling
CB3CB	T_m = 149.0 °C (ΔH = 30.9 kJ mol^{-1})	T_{INTB} = 49 °C [a]
CB2OCB	T_m = 139.7 °C (ΔH = 25.5 kJ mol^{-1})	T_{IN} = 108 °C [a], T_{NNTB} = 78 °C [a]
CB2SCB	T_m = 131.0 °C (ΔH = 24.4 kJ mol^{-1}) [b,c]	T_{INTB} = 42.6 °C [b]

[a] Determined by POM. [b] Obtained from Ref. [62]. [c] Obtained upon first heating.

Thus, CB3CB and CB2SCB exhibited the direct Iso–N_{TB} phase transition, whereas CB2OCB showed the typical Iso–N–N_{TB} phase transition. Moreover, the CB3CB and CB2OCB samples displayed a stronger crystallization tendency compared with the thioether-linked CB2SCB, which had a vitrifiable N_{TB} phase [62]. The Iso–LC phase-transition temperatures (i.e., T_{INTB} or T_{IN}) showed the order CB2OCB (108 °C) >> CB3CB (49 °C) > CB2SCB (43 °C), which translated to the order ether >> methylene > thioether in terms of linkage type. Naturally, the N_{TB} phase-transition temperatures (i.e., T_{INTB} or T_{NNTB}) also showed the same order: ether >> methylene > thioether. This particularly high phase-transition temperature (especially for the LC–Iso or Iso–LC phase transition) with the ether linkage is typical for usual calamitic LCs [69–71], including bent LC dimers [28,30,31,33,37,39]. Additionally, CB2SCB is vitrifiable, whereas CB3CB and CB2OCB strongly crystallize. These trends in the LC phase-transition temperatures and crystallization or vitrification abilities of the three dimers could be attributed to their different linkages, that is, ether (C–O–C), methylene (C–CH$_2$–C), and thioether (C–S–C). The C–O–C bond angle (118°) is larger than those of the other linkage types, which renders CB2OCB more anisotropic [72]. In addition, The higher rotational barrier [73,74] and stronger electron-donating property of the Ph–O bond may contribute to the molecular rigidity and intermolecular interactions of CB2OCB, respectively. These could result in higher T_{IN} and T_{NNTB}, a larger ΔT_N, and, possibly, at least in part, a stronger crystallization tendency. The higher T_{INTB} of CB3CB compared with that of CB2SCB is attributed to its higher molecular anisotropy owing to the larger C–CH$_2$–C bond angle (~110°) compared with the C–S–C angle (~100°), as well as higher rigidity due to the higher rotational barrier of the C–CH$_2$ bond compared with that of the C–S bond [75]. The higher T_m and T_{IN} and stronger crystallization tendency of CB3CB compared with those of CB2SCB could also be attributed to the symmetric molecular structure of the former. A more bent, flexible C–S–C linkage and molecular asymmetry endow CB2SCB with lower phase-transition temperatures and a vitrification ability compared with CB3CB and CB2OCB [33,39,62].

3.2. Phase-Transition Behaviors of CBnCB, CBnOCB, and CBnSCB

As described in the Introduction, each bent CB-based dimer homolog series with the longer spacers CBnCB (n = 5, 7, 9, 11, and 13) [17,25], CBnOCB (n = 4, 6, 8, and 10) [24,25], and CBnSCB (n = 4, 6, 8, and 10) [62] exhibit the typical Iso–N–N_{TB} phase sequence. The

T_{IN}, T_{NNTB}, and ΔT_N of these homologous series are plotted in Figure 5a–c, respectively, as a function of the total number of atoms in the spacer chain lengths, i.e., n for CBnCB and $n + 1$ for CBnOCB and CBnSCB, including the O and S atoms. The T_{INTB} values of CB3CB and CB2SCB are included in the plots shown in Figure 5a,b.

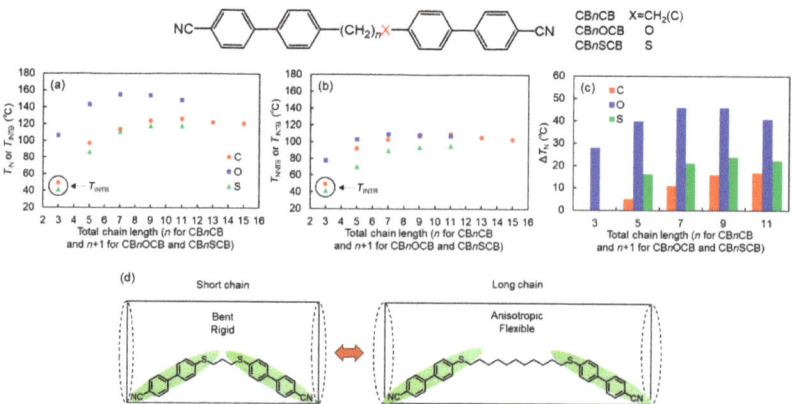

Figure 5. (a) T_{IN}, (b) T_{NNTB}, (c) ΔT_N, and (d) schematic models of dimers with shorter and longer spacers showing anisotropy and flexibility. In (a,b), the T_{INTB} values are plotted for CB3CB and CB2SCB because they exhibit the Iso–N$_{TB}$ phase transition. Panel (d) was reproduced from Ref. [39].

Overall, the T_{IN} and T_{NNTB} values were approximately in the order of CBnOCB > CBnCB > CBnSCB, which could be similarly ascribed to the characteristics of the linkers described in Section 3.1 for the shortest homologs. Nevertheless, the T_{NNTB} values of the ether-linked CBnOCB were relatively close to or partly lower than those of the methylene-linked CBnCB. This observation may partly be attributed to the characteristics of the more anisotropic structure of CBnOCB because the N$_{TB}$ phase formation for a more anisotropic molecular structure likely requires greater supercooling of the N phase compared to a more bent one [33]. Consequently, the ΔT_N values are in the order of CBnOCB > CBnSCB > CBnCB for all n, as shown in Figure 5c. The ΔT_N values of the ether-linked CBnOCB homologs are significantly larger than those of the others for all n owing to their high T_{IN} values.

Next, we investigated the n dependence of the phase-transition temperatures of the three homologs. The T_{IN} (or T_{INTB}) and T_{NNTB} values for all the CBnCB, CBnOCB, and CBnSCB homologs increase with increasing n and then level off or gradually decline with a further increase in n, as shown in Figure 5a,b. Consequently, with increasing n, ΔT_N increases for all the homologs, reaches a maximum, and then gradually declines for CBnCB and CBnOCB, as shown in Figure 5c. These trends of T_{IN} (or T_{INTB}), T_{NNTB}, and ΔT_N for all the dimer homologs could be attributed to the average molecular shape and flexibility with increasing/decreasing n, as shown in Figure 5d [39]. The shortest dimers could be more bent (strong biaxiality); hence, their T_{IN} (or T_{INTB}) and T_{NNTB} values were lower than those of the longer dimers. With increasing n, the average molecular shape of the bent dimer homologs becomes more linear (or anisotropic), thereby increasing T_{IN} and T_{NNTB} to some extent. However, further lengthening of the central spacer could enhance the molecular flexibility and dilute the polarizable mesogenic arms that increase the phase-transition temperatures; hence, these phase-transition temperatures nearly remain constant or gradually decline. Thus, the molecular biaxiality (molecular curvature) of the dimer homologs increases with decreasing n, which principally decreases T_{IN}, and consequently, ΔT_N. This results in $\Delta T_N = 0$, i.e., the direct Iso–N$_{TB}$ phase transition for the shortest CB3CB and CB2SCB [62].

4. Conclusions

In this study, we evaluated the phase-transition behaviors of three CB-based dimers with propane, ethoxy, and ethylthio spacers. Analogous to the previously reported CB2SCB, the short CB3CB exhibited the rare direct Iso–N_{TB} phase transition, whereas CB2OCB showed the typical Iso–N–N_{TB} phase transition. CB3CB and CB2OCB have strong crystallization tendencies, whereas the thioether-linked CB2SCB exhibited a vitrifiable N_{TB} phase. The N_{TB} phase-transition temperature (T_{INTB} or T_{NNTB}) decreased in the order CB2OCB (76 °C) > CB3CB (49 °C) > CB2SCB (43 °C). The phase-transition behaviors of all the CBnCB, CBnOCB, and CBnSCB homologs, including those with longer chains, were comprehensively examined. The more anisotropic ether-linked CBnOCB series showed significantly higher T_{IN} and wider ΔT_N for all n. Regarding shorter spacers, the phase-transition temperatures decreased, especially T_{IN}. Hence, the ΔT_N for all three homologous series decreased, resulting in the direct Iso–N_{TB} phase transition for the short-spacer-bearing CB3CB and CB2SCB. This phenomenon could partly be ascribed to their bent molecular geometry or enhanced molecular biaxiality owing to their short lengths. Our findings provide new insights into the effects of linkage types on the molecular design of LC dimers that exhibit the direct Iso–N_{TB} phase transition.

Author Contributions: Conceptualization, Y.A. (Yuki Arakawa); methodology, Y.A. (Yuki Arakawa); validation, Y.A. (Yuki Arakawa); formal analysis, Y.A. (Yuki Arakawa), Y.A. (Yuto Arai), K.H., and K.K.; investigation, Y.A. (Yuki Arakawa), Y.A. (Yuto Arai), K.H., and K.K.; resources, Y.A. (Yuki Arakawa) and H.T.; data curation, Y.A. (Yuki Arakawa); writing—original draft preparation, Y.A. (Yuki Arakawa); writing—review and editing, Y.A. (Yuki Arakawa), Y.A. (Yuto Arai), K.H., and H.T.; supervision, Y.A. (Yuki Arakawa) and H.T.; project administration, Y.A. (Yuki Arakawa); funding acquisition, Y.A. (Yuki Arakawa). All authors have read and agreed to the published version of the manuscript.

Funding: This research was funded by the Japan Society for the Promotion of Science (KAKENHI grant numbers 17K14493 and 20K15351) and Toyohashi University of Technology.

Data Availability Statement: Data are presented in the article.

Conflicts of Interest: The authors declare no financial conflict of interest.

References

1. Meyer, R.B. Structural Problems in Liquid Crystal Physics. In *Molecular Fluids: Summer School in Theoretical Physics, Les Houches Lectures 1973*; Balian, R., Weil, G., Eds.; Gordon and Breach: New York, NY, USA, 1976; pp. 271–343.
2. Dozov, I. On the spontaneous symmetry breaking in the mesophases of achiral banana-shaped molecules. *Europhys. Lett.* **2011**, *56*, 247–253. [CrossRef]
3. Memmer, R. Liquid crystal phases of achiral banana-shaped molecules: A computer simulation study. *Liq. Cryst.* **2002**, *29*, 483–496. [CrossRef]
4. Cestari, M.; Diez-Berart, S.; Dunmur, D.A.; Ferrarini, A.; de la Fuente, M.R.; Jackson, D.J.B.; Lopez, D.O.; Luckhurst, G.R.; Perez-Jubindo, M.A.; Richardson, R.M.; et al. Phase behavior and properties of the liquid-crystal dimer 1″,7″-bis (4-cyanobiphenyl-4′-yl) heptane: A twist-bend nematic liquid crystal. *Phys. Rev. E* **2011**, *84*, 31704. [CrossRef]
5. Panov, V.P.; Nagaraj, M.; Vij, J.K.; Panarin, Y.P.; Kohlmeier, A.; Tamba, M.G.; Lewis, R.A.; Mehl, G.H. Spontaneous periodic deformations in nonchiral planar-aligned bimesogens with a nematic-nematic transition and a negative elastic constant. *Phys. Rev. Lett.* **2010**, *105*, 167801. [CrossRef] [PubMed]
6. Chen, D.; Porada, J.H.; Hooper, J.B.; Klittnick, A.; Shen, Y.; Tuchband, M.R.; Korblova, E.; Bedrov, D.; Walba, D.M.; Glaser, M.A.; et al. Chiral heliconical ground state of nanoscale pitch in a nematic liquid crystal of achiral molecular dimers. *Proc. Natl. Acad. Sci. USA* **2013**, *110*, 15931–15936. [CrossRef]
7. Borshch, V.; Kim, Y.-K.; Xiang, J.; Gao, M.; Jákli, A.; Panov, V.P.; Vij, J.K.; Imrie, C.T.; Tamba, M.G.; Mehl, G.H.; et al. Nematic twist-bend phase with nanoscale modulation of molecular orientation. *Nat. Commun.* **2013**, *4*, 2635. [CrossRef]
8. Zhu, C.; Tuchband, M.R.; Young, A.; Shuai, M.; Scarbrough, A.; Walba, D.M.; Maclennan, J.E.; Wang, C.; Hexemer, A.; Clark, N.A. Resonant carbon K-edge soft X-ray scattering from lattice-free heliconical molecular ordering: Soft dilative elasticity of the twist-bend liquid crystal phase. *Phys. Rev. Lett.* **2016**, *116*, 147803. [CrossRef]
9. Salili, S.M.; Kim, C.; Sprunt, S.; Gleeson, J.T.; Parri, O.; Jákli, A. Flow properties of a twist-bend nematic liquid crystal. *RSC Adv.* **2014**, *4*, 57419–57423. [CrossRef]
10. Challa, P.K.; Borshch, V.; Parri, O.; Imrie, C.T.; Sprunt, S.N.; Gleeson, J.T.; Lavrentovich, O.D.; Jákli, A. Twist-bend nematic liquid crystals in high magnetic fields. *Phys. Rev. E* **2014**, *89*, 060501. [CrossRef]

11. Merkel, K.; Kocot, A.; Welch, C.; Mehl, G.H. Soft modes of the dielectric response in the twist–bend nematic phase and identification of the transition to a nematic splay bend phase in the CBC7CB dimer. *Phys. Chem. Chem. Phys.* **2019**, *21*, 22839–22848. [CrossRef] [PubMed]
12. Zhou, J.; Tang, W.; Arakawa, Y.; Tsuji, H.; Aya, S. Viscoelastic properties of a thioether-based heliconical twist–bend nematogen. *Phys. Chem. Chem. Phys.* **2020**, *22*, 9593–9599. [CrossRef] [PubMed]
13. Kumar, M.P.; Kula, P.; Dhara, S. Smecticlike rheology and pseudolayer compression elastic constant of a twist-bend nematic liquid crystal. *Phys. Rev. Mater.* **2020**, *4*, 115601. [CrossRef]
14. Vanakaras, A.G.; Photinos, D.J. A molecular theory of nematic–nematic phase transitions in mesogenic dimers. *Soft Matter* **2016**, *12*, 2208–2220. [CrossRef] [PubMed]
15. Samulski, E.T.; Reyes-Arango, D.; Vanakaras, A.G.; Photinos, D.J. All Structures Great and Small: Nanoscale Modulations in Nematic Liquid Crystals. *Nanomaterials* **2021**, *12*, 93. [CrossRef] [PubMed]
16. Sepelj, M.; Baumeister, U.; Ivšić, T.; Lesac, A. Effects of geometry and electronic structure on the molecular self-assembly of naphthyl-based dimers. *J. Phys. Chem. B* **2013**, *117*, 8918–8929. [CrossRef]
17. Henderson, P.A.; Imrie, C.T. Methylene-linked liquid crystal dimers and the twist-bend nematic phase. *Liq. Cryst.* **2011**, *38*, 1407–1414. [CrossRef]
18. Sebastian, N.; Lopez, D.O.; Robles-Hernandez, B.; de la Fuente, M.R.; Salud, J.; Perez-Jubindo, M.A.; Dunmur, D.A.; Luckhurst, G.R.; Jackson, D.J.B. Dielectric, calorimetric and mesophase properties of 1''-(2',4-difluorobiphenyl-4'-yloxy)-9''-(4-cyanobiphenyl-4'-yloxy) nonane: An odd liquid crystal dimer with a monotropic mesophase having the characteristics of a twist-bend nematic phase. *Phys. Chem. Chem. Phys.* **2014**, *16*, 21391–21406. [CrossRef]
19. Mandle, R.J.; Davis, E.J.; Archbold, C.T.; Voll, C.C.A.; Andrews, J.L.; Cowling, S.J.; Goodby, J.W. Apolar bimesogens and the incidence of the twist-bend nematic phase. *Chem. Eur. J.* **2015**, *21*, 8158–8167. [CrossRef]
20. Ahmed, Z.; Welch, C.; Mehl, G.H. The design and investigation of the selfassembly of dimers with two nematic phases. *RSC Adv.* **2015**, *5*, 93513–93521. [CrossRef]
21. Mandle, R.J.; Voll, C.C.A.; Lewis, D.J.; Goodby, J.W. Etheric bimesogens and the twist-bend nematic phase. *Liq. Cryst.* **2016**, *4*, 13–21. [CrossRef]
22. Mandle, R.J.; Goodby, J.W. Does Topology Dictate the Incidence of the Twist-Bend Phase? Insights Gained from Novel Unsymmetrical Bimesogens. *Chem. Eur. J.* **2016**, *22*, 18456–18464. [CrossRef]
23. Ivšić, T.; Vinković, M.; Baumeister, U.; Mikleušević, A.; Lesac, A. Towards understanding the N_{TB} phase: A combined experimental, computational and spectroscopic study. *RSC Adv.* **2016**, *6*, 5000–5007. [CrossRef]
24. Paterson, D.A.; Gao, M.; Kim, Y.K.; Jamali, A.; Finley, K.L.; Robles-Hernández, B.; Diez-Berart, S.; Salud, J.; de la Fuente, M.R.; Timimi, B.A.; et al. Understanding the twist-bend nematic phase: The characterisation of 1-(4-cyanobiphenyl-4'-yloxy)-6-(4-cyanobiphenyl-4'-yl)hexane (CB6OCB) and comparison with CB7CB. *Soft Matter* **2016**, *12*, 6827–6840. [CrossRef] [PubMed]
25. Paterson, D.A.; Abberley, J.P.; Harrison, W.T.A.; Storey, J.M.D.; Imrie, C.T. Cyanobiphenyl-based liquid crystal dimers and the twist-bend nematic phase. *Liq. Cryst.* **2017**, *44*, 127–146. [CrossRef]
26. Knežević, A.; Sapunar, M.; Buljan, A.; Dokli, I.; Hameršak, Z.; Kontrec, D.; Lesac, A. Fine-tuning the effect of pep interactions on the stability of the N_{TB} phase. *Soft Matter* **2018**, *14*, 8466–8474. [CrossRef] [PubMed]
27. Watanabe, K.; Tamura, T.; Kang, S.; Tokita, M. Twist bend nematic liquid crystals prepared by one-step condensation of 4-(4-Pentylcyclohexyl) benzoic acid and alkyl diol. *Liq. Cryst.* **2018**, *45*, 924–930. [CrossRef]
28. Arakawa, Y.; Komatsu, K.; Tsuji, H. Twist-bend nematic liquid crystals based on thioether linkage. *New J. Chem.* **2019**, *43*, 6786–6793. [CrossRef]
29. Zep, A.; Pruszkowska, K.; Dobrzycki, Ł.; Sektas, K.; Szałański, P.; Marek, P.H.; Cyrański, M.K.; Sicinski, R.R. Cholesterol-based photo-switchable mesogenic dimers. Strongly bent molecules versus an intercalated structure. *CrystEngComm* **2019**, *21*, 2779–2789. [CrossRef]
30. Arakawa, Y.; Tsuji, H. Selenium-linked liquid crystal dimers for twist-bend nematogens. *J. Mol. Liq.* **2019**, *289*, 111097. [CrossRef]
31. Cruickshank, E.; Salamonczyk, M.; Pociecha, D.; Strachan, G.J.; Storey, J.M.D.; Wang, C.; Feng, J.; Zhu, C.; Gorecka, E.; Imrie, C.T. Sulfur-linked cyanobiphenylbased liquid crystal dimers and the twist-bend nematic phase. *Liq. Cryst.* **2019**, *46*, 1595–1609. [CrossRef]
32. Walker, R.; Pociecha, D.; Storey, J.M.D.; Gorecka, E.; Imrie, C.T. The Chiral Twist-Bend Nematic Phase (N^*_{TB}). *Chem. Eur. J.* **2019**, *25*, 13329–13335. [CrossRef] [PubMed]
33. Arakawa, Y.; Ishida, Y.; Tsuji, H. Ether-and Thioether-Linked Naphthalene-Based Liquid-Crystal Dimers: Influence of Chalcogen Linkage and Mesogenic-Arm Symmetry on the Incidence and Stability of the Twist–Bend Nematic Phase. *Chem. Eur. J.* **2020**, *26*, 3767–3775. [CrossRef] [PubMed]
34. Forsyth, E.; Paterson, D.A.; Cruickshank, E.; Strachan, G.J.; Gorecka, E.; Walker, R.; Storey, J.M.D.; Imrie, C.T. Liquid crystal dimers and the twist-bend nematic phase: On the role of spacers and terminal alkyl chains. *J. Mol. Liq.* **2020**, *320*, 114391. [CrossRef]
35. Arakawa, Y.; Komatsu, K.; Feng, J.; Zhu, C.; Tsuji, H. Distinct twist-bend nematic phase behaviors associated with the ester-linkage direction of thioether-linked liquid crystal dimers. *Mater. Adv.* **2021**, *2*, 261–272. [CrossRef]
36. Walker, R.; Majewska, M.; Pociecha, D.; Makal, A.; Storey, J.M.D.; Gorecka, E.; T Imrie, C. Twist-bend nematic glasses: The synthesis and characterisation of pyrene-based nonsymmetric dimers. *ChemPhysChem* **2021**, *22*, 461–470. [CrossRef] [PubMed]

37. Arakawa, Y.; Ishida, Y.; Komatsu, K.; Arai, Y.; Tsuji, H. Thioether-linked benzylideneaniline-based twist-bend nematic liquid crystal dimers: Insights into spacer lengths, mesogenic arm structures, and linkage types. *Tetrahedron* **2021**, *95*, 132351. [CrossRef]
38. Arakawa, Y.; Komatsu, K.; Ishida, Y.; Igawa, K.; Tsuji, H. Carbonyl-and thioether-linked cyanobiphenyl-based liquid crystal dimers exhibiting twist-bend nematic phases. *Tetrahedron* **2021**, *81*, 131870. [CrossRef]
39. Arakawa, Y. Twist-bend Nematic Liquid Crystals Bearing Chalcogen-based Linkages. *EKISHO* **2021**, *25*, 221–229.
40. Mandle, R.J.; Goodby, J.W. A liquid crystalline oligomer exhibiting nematic and twist-bend nematic mesophases. *ChemPhysChem* **2016**, *17*, 967–970. [CrossRef] [PubMed]
41. Simpson, F.P.; Mandle, R.J.; Moore, J.N.; Goodby, J.W. Investigating the Cusp between the nano-and macro-sciences in supermolecular liquid-crystalline twist-bend nematogens. *J. Mater. Chem. C* **2017**, *5*, 5102–5110. [CrossRef]
42. Tuchband, M.R.; Paterson, D.A.; Salamończyk, M.; Norman, V.A.; Scarbrough, A.N.; Forsyth, E.; Garcia, E.; Wang, C.; Storey, J.M.D.; Walba, D.M.; et al. Distinct differences in the nanoscale behaviors of the twist–bend liquid crystal phase of a flexible linear trimer and homologous dimer. *Proc. Natl. Acad. Sci. USA* **2019**, *116*, 10698–10704. [CrossRef] [PubMed]
43. Saha, R.; Babakhanova, G.; Parsouzi, Z.; Rajabi, M.; Gyawali, P.; Welch, C.; Mehl, G.H.; Gleeson, J.; Lavrentovich, O.D.; Sprunt, S.; et al. Oligomeric odd-even effect in liquid crystals. *Mater. Horiz.* **2019**, *6*, 1905–1912. [CrossRef]
44. Arakawa, Y.; Komatsu, K.; Inui, S.; Tsuji, H. Thioether-linked liquid crystal dimers and trimers: The twist-bend nematic phase. *J. Mol. Struct.* **2020**, *1199*, 126913. [CrossRef]
45. Al-Janabi, A.; Mandle, R.J. Utilising saturated hydrocarbon isosteres of para benzene in the design of twist-bend nematic liquid crystals. *ChemPhysChem* **2020**, *21*, 697–701. [CrossRef] [PubMed]
46. Arakawa, Y.; Komatsu, K.; Shiba, T.; Tsuji, H. Phase behaviors of classic liquid crystal dimers and trimers: Alternate induction of smectic and twist-bend nematic phases depending on spacer parity for liquid crystal trimers. *J. Mol. Liq.* **2021**, *326*, 115319. [CrossRef]
47. Majewska, M.M.; Forsyth, E.; Pociecha, D.; Wang, C.; Storey, J.M.D.; Imrie, C.T.; Gorecka, E. Controlling spontaneous chirality in achiral materials: Liquid crystal oligomers and the heliconical twist-bend nematic phase. *Chem. Commun.* **2022**, *58*, 5285–5288. [CrossRef]
48. Arakawa, Y.; Komatsu, K.; Tsuji, H. 2, 7-substituted fluorenone-based liquid crystal trimers: Twist-bend nematic phase induced by outer thioether linkage. *Phase Transit.* **2022**, *95*, 331–339. [CrossRef]
49. Arakawa, Y.; Komatsu, K.; Ishida, Y.; Shiba, T.; Tsuji, H. Thioether-linked liquid crystal trimers: Odd–even effects of spacers and the influence of thioether bonds on phase behavior. *Materials* **2022**, *15*, 1709. [CrossRef]
50. Mandle, R.J.; Goodby, J.W. A Nanohelicoidal Nematic Liquid Crystal Formed by a Non-Linear Duplexed Hexamer. *Angew. Chem. Int. Ed.* **2018**, *57*, 7096–7100. [CrossRef]
51. Stevenson, W.D.; An, J.; Zeng, X.B.; Xue, M.; Zou, H.X.; Liu, Y.S.; Ungar, G. Twist-bend nematic phase in biphenylethane-based copolyethers. *Soft Matter* **2018**, *14*, 3003–3011. [CrossRef]
52. Jansze, S.M.; Martínez-Felipe, A.; Storey, J.M.D.; Marcelis, A.T.M.; Imrie, C.T. A twist-bend nematic phase driven by hydrogen bonding. *Angew. Chem. Int. Ed.* **2015**, *127*, 653–656. [CrossRef]
53. Walker, R.; Pociecha, D.; Martinez-Felipe, A.; Storey, J.M.D.; Gorecka, E.; Imrie, C.T. Twist-bend nematogenic supramolecular dimers and trimers formed by hydrogen bonding. *Crystals* **2020**, *10*, 175. [CrossRef]
54. Chen, D.; Nakata, M.; Shao, R.; Tuchband, M.R.; Shuai, M.; Baumeister, U.; Weissflog, W.; Walba, D.M.; Glaser, M.A.; Maclennan, J.E.; et al. Twist-bend heliconical chiral nematic liquid crystal phase of an achiral rigid bent-core mesogen. *Phys. Rev. E Stat. Nonlinear Biol. Soft Matter Phys.* **2014**, *89*, 022506. [CrossRef] [PubMed]
55. Wang, Y.; Singh, G.; Agra-Kooijman, D.M.; Gao, M.; Bisoyi, H.K.; Xue, C.; Fisch, M.R.; Kumar, S.; Li, Q. Room temperature heliconical twist-bend nematic liquid crystal. *CrystEngComm* **2015**, *17*, 2778–2782. [CrossRef]
56. Sreenilayam, S.P.; Panov, V.P.; Vij, J.K.; Shanker, G. The N_{TB} phase in an achiral asymmetrical bent-core liquid crystal terminated with symmetric alkyl chains. *Liq. Cryst.* **2017**, *44*, 244–253. [CrossRef]
57. Mandle, R.J. A Ten-Year Perspective on Twist-Bend Nematic Materials. *Molecules* **2022**, *27*, 2689. [CrossRef]
58. Mandle, R.J.; Archbold, C.T.; Sarju, J.P.; Andrews, J.L.; Goodby, J.W. The dependency of nematic and twist-bend mesophase formation on bend angle. *Sci. Rep.* **2016**, *6*, 36682. [CrossRef]
59. Mandle, R.J.; Goodby, J.W. Molecular flexibility and bend in semi-rigid liquid crystals: Implications for the heliconical nematic ground state. *Chem. Eur. J.* **2019**, *25*, 14454. [CrossRef]
60. Yu, G.; Wilson, M.R. All-atom simulations of bent liquid crystal dimers: The twist-bend nematic phase and insights into conformational chirality. *Soft Matter* **2022**, *18*, 3087–3096. [CrossRef]
61. Kumar, A. Dependency of the twist-bend nematic phase formation on the molecular shape of liquid crystal dimers: A view through the lens of DFT. *J. Mol. Liq.* **2022**, *354*, 118858. [CrossRef]
62. Arakawa, Y.; Komatsu, K.; Shiba, T.; Tsuji, H. Methylene-and thioether-linked cyanobiphenyl-based liquid crystal dimers CBnSCB exhibiting room temperature twist-bend nematic phases and glasses. *Mater. Adv.* **2021**, *2*, 1760–1773. [CrossRef]
63. Archbold, C.T.; Davis, E.J.; Mandle, R.J.; Cowling, S.J.; Goodby, J.W. Chiral dopants and the twist-bend nematic phase–induction of novel mesomorphic behaviour in an apolar bimesogen. *Soft Matter* **2015**, *11*, 7547–7557. [CrossRef]
64. Dawood, A.A.; Grossel, M.C.; Luckhurst, G.R.; Richardson, R.M.; Timimi, B.A.; Wells, N.J.; Yousif, Y.Z. On the twist-bend nematic phase formed directly from the isotropic phase. *Liq. Cryst.* **2016**, *43*, 2–12. [CrossRef]

65. Dawood, A.A.; Grossel, M.C.; Luckhurst, G.R.; Richardson, R.M.; Timimi, B.A.; Wells, N.J.; Yousif, Y.Z. Twist-bend nematics, liquid crystal dimers, structure–property relations. *Liq. Cryst.* **2017**, *44*, 106–126.
66. Wang, D.; Liu, J.; Zhao, W.; Zeng, Y.; Huang, J.; Fang, J.; Chen, D. Facile Synthesis of Liquid Crystal Dimers Bridged with a Phosphonic Group. *Chem. Eur. J.* **2022**. [CrossRef]
67. Ramou, E.; Welch, C.; Hussey, J.; Ahmed, Z.; Karahaliou, P.K.; Mehl, G.H. The induction of the N_{tb} phase in mixtures. *Liq. Cryst.* **2018**, *45*, 1929–1935. [CrossRef]
68. Akdag, A.; Wahab, A.; Beran, P.; Rulisek, L.; Dron, P.I.; Ludvik, J.; Michl, J. Covalent Dimers of 1,3-Diphenylisobenzofuran for Singlet Fission: Synthesis and Electrochemistry. *J. Org. Chem.* **2015**, *80*, 80–89. [CrossRef]
69. Arakawa, Y.; Kang, S.; Tsuji, H.; Watanabe, J.; Konishi, G. Development of novel bistolane-based liquid crystalline molecules with an alkylsulfanyl group for highly birefringent materials. *RSC Adv.* **2016**, *6*, 16568–16574. [CrossRef]
70. Arakawa, Y.; Tsuji, H. Phase transitions and birefringence of bistolane-based nematic molecules with an alkyl, alkoxy and alkylthio group. *Mol. Cryst. Liq. Cryst.* **2017**, *647*, 422–429. [CrossRef]
71. Arakawa, Y.; Sasaki, Y.; Tsuji, H. Supramolecular hydrogen-bonded liquid crystals based on 4-*n*-alkylthiobenzoic acids and 4,4′-bipyridine: Their mesomorphic behavior with comparative study including alkyl and alkoxy counterparts. *J. Mol. Liq.* **2019**, *280*, 153–159. [CrossRef]
72. Emerson, A.P.J.; Luckhurst, G.R. On the relative propensities of ether and methylene linkages for liquid crystal formation in calamitics. *Liq. Cryst.* **1991**, *10*, 861–868. [CrossRef]
73. Tasaka, T.; Okamoto, H.; Petrov, V.F.; Takenaka, S. Liquid crystalline properties of dissymmetric molecules part 5: The effects of alkyl chain length and linkages on thermal properties of smectic A and C phases in three aromatic ring systems. *Mol. Cryst. Liq. Cryst.* **2001**, *357*, 67–84. [CrossRef]
74. Balema, T.A.; Ulumuddin, N.; Murphy, C.J.; Slough, D.P.; Smith, Z.C.; Hannagan, R.T.; Wasio, N.A.; Larson, A.M.; Patel, D.A.; Groden, K.; et al. Controlling molecular switching via chemical functionality: Ethyl vs methoxy rotors. *J. Phys. Chem. C* **2019**, *123*, 23738–23746. [CrossRef]
75. McCurdy, R.M.; Prager, J.H. Thiaalkyl polyacrylates: The influence of sulfur in the side chain. *J. Polym. Sci. Part A Gen. Pap.* **1964**, *2*, 1185–1192. [CrossRef]

Article

Development of Hydrogen-Bonded Dimer-Type Photoluminescent Liquid Crystals of Fluorinated Tolanecarboxylic Acid

Shigeyuki Yamada [1,*,†], Mitsuki Kataoka [1,†], Keigo Yoshida [1,†], Masakazu Nagata [2,‡], Tomohiro Agou [2], Hiroki Fukumoto [2] and Tsutomu Konno [1]

[1] Faculty of Molecular Chemistry and Engineering, Kyoto Institute of Technology, Matsugasaki, Sakyo-ku, Kyoto 606-8585, Japan
[2] Department of Quantum Beam Science, Graduate School of Science and Engineering, Ibaraki University, 4-12-1 Nakanarusawa, Hitachi 316-8511, Ibaraki, Japan
* Correspondence: syamada@kit.ac.jp; Tel.: +81-75-724-7517
[†] The authors contributed equally to this work.
[‡] Current address: Organic Materials Chemistry Group, Sagami Chemical Research Institute, Hayakawa 2743-1, Ayase 252-1193, Kanagawa, Japan.

Abstract: Functional molecules possessing photoluminescence (PL) and liquid-crystalline (LC) behaviors, known as photoluminescent liquid crystals, along with a small molecular structure, have attracted significant attention. Fluorinated tolane skeletons are small π-conjugated structures, which are promising candidates for such functional molecules. These structures were revealed to exhibit strong PL in solid state but no LC behavior. Based on a report on hydrogen-bonded dimer-type LC molecules of carboxylic acid, in this study, we designed and synthesized a series of fluorinated tolanecarboxylic acids (2,3,5,6-tetrafluoro-4-[2-(4-alkoxyphenyl)ethyn-1-yl]benzoic acids) as promising PLLC molecules. Evaluation of the LC behavior revealed that fluorinated tolanecarboxylic acids with a longer alkoxy chain than a butoxy chain exhibited nematic LC behavior. Additionally, fluorinated tolanecarboxylic acids showed intense PL in the solution and crystalline states. Notably, fluorinated tolanecarboxylic acid with an aggregated structure in the nematic LC phase also exhibited PL with a slight blue shift in PL maximum wavelength compared to the crystalline state. The present fluorinated tolanecarboxylic acid exhibiting PL and LC characteristics in a single molecule can be applied to thermoresponsive PL materials, such as a PL thermosensor.

Keywords: diphenylacetylene; fluorinated tolanecarboxylic acid; fluorine; photoluminescence; liquid crystals; nematic phase; phase transition

Citation: Yamada, S.; Kataoka, M.; Yoshida, K.; Nagata, M.; Agou, T.; Fukumoto, H.; Konno, T. Development of Hydrogen-Bonded Dimer-Type Photoluminescent Liquid Crystals of Fluorinated Tolanecarboxylic Acid. *Crystals* **2023**, *13*, 25. https://doi.org/10.3390/cryst13010025

Academic Editors: Ingo Dierking, Charles Rosenblatt, Kyosuke Isoda, Takahiro Ichikawa, Kosuke Kaneko, Mizuho Kondo, Tsuneaki Sakurai, Atsushi Seki, Mitsuo Hara and Go Watanabe

Received: 24 November 2022
Revised: 19 December 2022
Accepted: 21 December 2022
Published: 23 December 2022

Copyright: © 2022 by the authors. Licensee MDPI, Basel, Switzerland. This article is an open access article distributed under the terms and conditions of the Creative Commons Attribution (CC BY) license (https:// creativecommons.org/licenses/by/ 4.0/).

1. Introduction

Photoluminescent liquid crystals, which possess photoluminescence (PL) and liquid-crystalline (LC) characteristics in a single molecule, have gained recognition as essential organic functional molecules owing to their extensive applicability in PL thermometers and thermoresponsive PL sensors [1–3]. To date, many PLLC molecules have been developed [4,5], which consist of large molecular structures with a π-conjugated structure, mesogenic core, and flexible unit that result in PL and LC behaviors. Therefore, developing PLLC molecules with a small molecular structure is necessary for practical applications considering the manufacturing costs and processes. An effective approach to searching for PLLC molecules with a small molecular structure is designing a common π-conjugated structure that functions as the core structure of PL and LC molecules.

Over the past few years, our group has focused on developing fluorine-containing organic functional molecules with a PL and an LC characteristic [6–14]. Our recent study revealed that fluorinated bistolane-based PLLC molecules (**A**) exhibit PL and LC behaviors in a single molecule. The PL behavior is switched depending on the structural changes in the molecular aggregates through phase transition between the crystalline (Cry) and LC phases (Figure 1a) [6,7]. However, several issues were to be resolved, thus requiring multiple reaction steps to synthesize bistolane-based PLLC molecules. Because fluorinated tolane derivatives exhibit intense PL in the Cry phase through intermolecular H···F hydrogen bonds [8–11], we suggested that a fluorinated tolane skeleton, which contains a small and common π-conjugated structure, is effective as the core structure of the PL and LC molecules. Several attempts revealed that alkoxy-substituted fluorinated tolanes with a cyano (CN) [8], a trifluoromethyl (CF_3) group [8], and a fluorine (F) atom [9] show intense PL in the Cry phase but no LC phase, whereas fluorinated tolanecarboxylates **B** with a long flexible alkoxy chain, such as $C_7H_{15}O$ and $C_8H_{17}O$, reportedly exhibit intense PL in the Cry phase and the nematic (N) LC phase after the cooling process (Figure 1b) [12]. Additionally, fluorinated tolane dimer **C**, which is composed of two fluorinated tolane skeletons connected by a flexible chain, successfully exhibits the PL and LC phases (Figure 1c) [13,14].

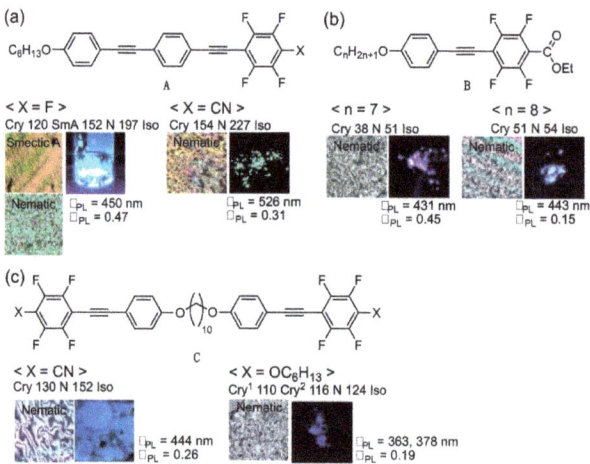

Figure 1. Chemical structure, phase transition behavior, and crystalline-state photoluminescence (PL) behavior of (**a**) fluorinated bistolane-based PL liquid crystals (PLLCs) (**A**), (**b**) fluorinated tolane-based PLLCs (**B**), and (**c**) fluorinated tolane dimer-type PLLCs (**C**).

Arakawa et al. reported that aromatic carboxylic acids, including tolanecarboxylic acid, show broad LC behavior due to formation of dimer via hydrogen bonds [15,16]. Wen et al. examined fluorinated LC molecules and reported that fluorinated tolanes with an ester structure [17] or fluorinated biphenyls with a carboxy unit exhibit LC behavior [18]. Based on the molecular design of hydrogen-bonded dimer-type LC molecules, we focused on the hydrogen-bonded dimer-type LC molecules of carboxylic acid. In this study, we designed and synthesized a series of fluorinated tolanecarboxylic acids **1**, such as 2,3,5,6-tetrafluoro-4-[2-(4-alkoxyphenyl)ethyn-1-yl]benzoic acids (Figure 2), and evaluated their LC and PL characteristics in detail.

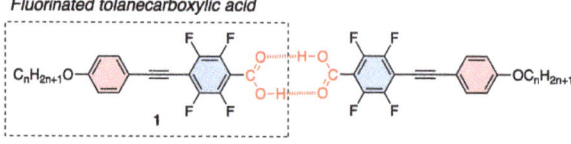

Figure 2. Chemical structure of the fluorinated tolanecarboxylic acid **1** used in this study and plausible aggregated structure in crystalline and LC phases through hydrogen bond.

2. Materials and Methods

2.1. General

Column chromatography was performed for purification using Wakogel® 60N (38–100 μm), and thin layer chromatography (TLC) analysis was performed on silica gel TLC plates (silica gel 60F$_{254}$, Merck). The melting temperature (T_m) and clearing temperature (T_c) were determined using polarized optical microscopy (POM). ^{1}H and ^{13}C nuclear magnetic resonance (NMR) spectra were obtained using a Bruker AVANCE III 400 NMR spectrometer (^{1}H: 400 MHz and ^{13}C: 100 MHz) in chloroform-d (CDCl$_3$) or dimethyl sulfoxide-d_6 or acetone-d_6, and chemical shifts were reported in parts per million (ppm) using the residual proton in the NMR solvent. ^{19}F NMR (376 MHz) spectra were obtained using a Bruker AVANCE III 400 NMR spectrometer in CDCl$_3$; CFCl$_3$ (δ_F = 0.0 ppm) and hexafluorobenzene (δ_F = −163 ppm) were used as internal standards. Infrared (IR) spectra were recorded using the KBr method with a JASCO FT/IR-4100 type A spectrometer; all spectra were reported in wavenumber (cm^{-1}) unit. High-resolution mass spectrometry (HRMS) was performed on a JEOL JMS-700MS spectrometer using the fast atom bombardment (FAB) method. Synthetic precursor ethyl 4-[2-(4-alkoxyphenyl)ethyn-1-yl]-2,3,5,6-tetrafluorobenzoate (**2**) was stated in a previous study and synthesized according to the reported procedure [12].

2.2. Typical Synthetic Procedure of 2,3,5,6-Tetrafluoro-2-[4-(methoxyphenyl)ethyn-1-yl]benzoic acid (**1a**)

Ethyl 2,3,5,6-tetrafluoro-4-[2-(4-methoxyphenyl)ethyn-1-yl]benzoate (**2a**, 2.0 g, 5.7 mmol), tetrahydrofuran (THF, 28 mL), and H$_2$O (12 mL) were placed in a two-necked round-bottomed flask, followed by addition of LiOH·H$_2$O (0.6 g, 14 mmol). The mixture was stirred at room temperature for 20 h and then acidified by adding an aqueous solution of HCl until the pH of the solution was below 1. The crude product was extracted with Et$_2$O (10 mL, three times), while the organic layer was washed with brine (20 mL, once). The collected organic layer was dried over anhydrous Na$_2$SO$_4$ and separated from the drying agent by atmospheric filtration. The filtrate was evaporated using a rotary evaporator under reduced pressure and subjected to column chromatography using hexane, ethyl acetate, and acetic acid ($v/v/v$ = 50/50/1) as an eluent, followed by recrystallization from chloroform, generating the title molecule **1a** as a white solid in a 74% isolated yield (1.37 g, 4.2 mmol).

2.2.1. 2,3,5,6-Tetrafluoro-4-{2-(4-methoxyphenyl)ethyn-1-yl}benzoic acid (**1a**)

Yield: 74% (white solid); T_m: 223 °C (determined by POM); ^{1}H NMR (DMSO-d_6): δ 3.82 (s, 3H), 7.05 (d, J = 8.8 Hz, 2H), 7.59 (d, J = 8.8 Hz, 2H), 14.55 (brs, 1H); ^{13}C NMR (DMSO-d_6): δ 55.4, 72.8 (t, J = 4.4 Hz), 103.5 (t, J = 3.6 Hz), 105.9 (t, J = 17.6 Hz), 111.9, 113.6 (t, J = 17.6 Hz), 114.7, 133.6, 143.7 (dm, J = 253.1 Hz), 146.0 (dm, J = 250.8 Hz), 160.0, 160.8; ^{19}F NMR (DMSO-d_6, CFCl$_3$): δ −136.3 to −136.6 (m, 2F), −138.9 to −139.1 (m, 2F); IR (KBr): v 3730, 2844, 2221, 1698, 1601, 1475, 1247, 1174, 990, 835 cm^{-1}; HRMS (FAB): [M+] calcd C$_{16}$H$_8$F$_4$O$_3$: 324.0410, found: 324.0413.

2.2.2. 2-{(4-Ethoxyphenyl)ethyn-1-yl}-2,3,5,6-tetrafluorobenzoic acid (**1b**)

Yield: 89% (white solid); T_m: 224 °C (determined by POM); ^{1}H NMR (acetone-d_6): δ 1.39 (t, J = 7.2 Hz, 3H), 4.12 (q, J = 7.2 Hz, 2H), 7.02 (d, J = 8.8 Hz, 2H), 7.58 (d, J = 8.8 Hz, 2H), 6.0–8.0 (brs, 1H); ^{13}C NMR (acetone-d_6): δ 14.9, 64.5, 73.4 (t, J = 3.6 Hz), 105.0 (t, J = 3.7 Hz), 108.1 (t, J = 17.6 Hz), 113.5, 113.6 (t, J = 17.6 Hz), 115.9, 134.5, 145.5 (ddt, J = 253.8, 13.2, 5.9 Hz), 147.4 (ddt, J = 250.9, 14.7, 3.6 Hz), 160.2, 161.6; ^{19}F NMR (acetone-d_6, C$_6$F$_6$): δ −136.92 (dd, J = 20.7, 10.9 Hz, 2F), −140.31 (dd, J = 20.7, 10.9 Hz, 2F); IR (KBr): v 3750, 2984, 2212, 1706, 1601, 1479, 1178, 994, 844 cm^{-1}; HRMS (FAB): [M+] calcd C$_{17}$H$_{10}$F$_4$O$_3$: 338.0566, found: 338.0563.

2.2.3. 2,3,5,6-Tetrafluoro-4-{2-(4-propyloxy)ethyn-1-yl}benzoic acid (1c)

Yield: 83% (white solid); T_m: 220 °C (determined by POM); ^{1}H NMR (acetone-d_6): δ 1.03 (t, J = 7.2 Hz, 3H), 1.81 (sext., J = 7.2 Hz, 2H), 4.031 (t, J = 6.8 Hz, 2H), 7.04 (d, J = 8.8 Hz, 2H), 7.58 (d, J = 8.8 Hz, 2H), 5.0–10 (brs, 1H); ^{13}C NMR (acetone-d_6): δ 10.7, 23.1, 70.4, 73.4 (t, J = 4.4 Hz), 105.0 (t, J = 3.6 Hz), 108.1 (t, J = 17.6 Hz), 113.5, 113.8 (t, J = 16.9 Hz), 115.9, 134.5, 145.4 (ddt, J = 253.8, 13.9, 5.1 Hz), 147.4 (ddt, J = 250.9, 15.4, 3.6 Hz), 160.3, 161.8; ^{19}F NMR (acetone-d_6, C_6F_6): δ −136.91 (dd, J = 20.7, 10.5 Hz, 2F), −140.3 (dd, J = 20.7, 10.5 Hz, 2F); IR (KBr): ν 3650, 2966, 2211, 1705, 1601, 1476, 1331, 1253, 993 cm^{-1}; HRMS (FAB): [M+] calcd $C_{18}H_{12}F_4O_3$: 352.0723, found: 352.0733.

2.2.4. 2-{(4-Butoxyphenyl)ethyn-1-yl}-2,3,5,6-tetrafluorobenzoic acid (1d)

Yield: 78% (white solid); T_m: 178 °C (determined by POM); ^{1}H NMR (acetone-d_6): δ 0.97 (t, J = 7.2 Hz, 3H), 1.50 (sext., J = 7.2 Hz, 2H), 1.76 (quin, J = 7.2 Hz, 2H), 4.03 (t, J = 6.8 Hz, 2H), 6.98 (d, J = 8.8 Hz, 2H), 7.52 (d, J = 8.8 Hz, 2H), 10.0 (brs, 1H); ^{13}C NMR (CDCl$_3$): δ 14.1, 19.8, 31.9, 68.6, 73.4 (t, J = 5.1 Hz), 105.0 (t, J = 3.7 Hz), 108.1 (t, J = 16.1 Hz), 113.5, 113.5 (t, J = 16.2 Hz), 115.8, 134.4, 145.5 (ddt, J = 252.3, 13.2, 5.8 Hz), 147.3 (ddt, J = 253.0, 13.9, 3.6 Hz), 160.3, 161.7; ^{19}F NMR (acetone-d_6, C_6F_6): δ −136.9 to −137.1 (m, 2F), −140.2 to −140.4 (m, 2F); IR (KBr): ν 3743, 2950, 2209, 1707, 1600, 1477, 1252, 1177, 993 cm^{-1}; HRMS (FAB): [M+] calcd $C_{19}H_{14}F_4O_3$: 366.0879, found: 366.0893.

2.2.5. 2,3,5,6-Tetrafluoro-4-{2-(4-pentyloxy)ethyn-1-yl}benzoic acid (1e)

Yield: 44% (white solid); T_m: 175 °C (determined by POM); ^{1}H NMR (acetone-d_6): δ 0.94 (t, J = 7.2 Hz, 3H), 1.35–1.52 (m, 4H), 1.80 (quin, J = 6.8 Hz, 2H), 4.07 (t, J = 6.8 Hz, 2H), 7.01 (d, J = 8.8 Hz, 2H), 7.55 (d, J = 8.8 Hz, 2H), 9.07 (brs, 1H); ^{13}C NMR (acetone-d_6): δ 14.3, 23.1, 29.0, 29.7, 69.2, 73.5 (t, J = 4.4 Hz), 105.2 (t, J = 3.6 Hz), 108.4 (t, J = 18.3 Hz), 113.8, 113.9 (t, J = 17.6 Hz), 116.1, 134.6, 145.6 (ddt, J = 255.2, 15.3, 4.4 Hz), 147.6 (ddt, J = 250.9, 14.7, 3.6 Hz), 160.3, 162.0; ^{19}F NMR (acetone-d_6, C_6F_6): δ −136.92 (dd, J = 20.7, 10.9 Hz, 2F), −140.28 (dd, J = 20.7, 10.9 Hz, 2F); IR (KBr): ν 3485, 2948, 2212, 1707, 1600, 1481, 1253, 1176, 996, 837 cm^{-1}; HRMS (FAB): [M+] calcd $C_{20}H_{16}F_4O_3$: 380.1036, found: 380.1027.

2.2.6. 2,3,5,6-Tetrafluoro-4-{2-(4-hexyloxy)ethyn-1-yl}benzoic acid (1f)

Yield: 65% (white solid); T_m: 185 °C (determined by POM); ^{1}H NMR (acetone-d_6): δ 0.90 (t, J = 6.8 Hz, 3H), 1.30–1.38 (m, 4H), 1.48 (quin, J = 6.8 Hz, 2H), 1.79 (quin, J = 6.8 Hz, 2H), 4.06 (t, J = 6.8 Hz, 2H), 7.02 (d, J = 8.8 Hz, 2H), 7.57 (d, J = 8.8 Hz, 2H), 8.94 (brs, 1H); ^{13}C NMR (acetone-d_6): δ 14.3, 23.3, 26.4, 32.3, 68.9, 73.4 (t, J = 3.7 Hz), 105.0 (t, J = 3.6 Hz), 108.1 (t, J = 18.4 Hz), 113.5, 113.6 (t, J = 16.9 Hz), 115.9, 134.5, 145.5 (ddt, J = 252.3, 14.0, 5.1 Hz), 147.4 (ddt, J = 251.5, 14.7, 3.7 Hz), 160.2, 161.8; ^{19}F NMR (acetone-d_6, C_6F_6): δ −136.97 (dd, J = 20.3, 12.4 Hz, 2F), −140.29 (dd, J = 20.3, 12.4 Hz, 2F); IR (KBr): ν 3450, 2946, 2211, 1705, 1602, 1476, 1329, 1172, 993, 834 cm^{-1}; HRMS (FAB): [M+] calcd $C_{21}H_{18}F_4O_3$: 394.1192, found: 394.1202.

2.2.7. 2,3,5,6-Tetrafluoro-4-{2-(4-heptyloxy)ethyn-1-yl}benzoic acid (1g)

Yield: 75% (white solid); T_m: 172 °C (determined by POM); ^{1}H NMR (acetone-d_6): δ 0.89 (t, J = 6.8 Hz, 3H), 1.26–1.42 (m, 6H), 1.48 (quin, J = 6.8 Hz, 2H), 1.80 (quin, J = 6.8 Hz, 2H), 4.07 (t, J = 6.8 Hz, 2H), 7.03 (d, J = 8.8 Hz, 2H), 7.58 (d, J = 8.8 Hz, 2H), 6.0–10 (brs, 1H); ^{13}C NMR (acetone-d_6): δ 14.3, 23.3, 26.7, 29.8, 29.9, 32.6, 68.9, 73.4 (t, J = 4.4 Hz), 105.0 (t, J = 3.7 Hz), 108.1 (t, J = 17.6 Hz), 113.5, 113.7 (t, J = 16.9 Hz), 115.9, 134.5, 145.5 (ddt, J = 253.1, 13.9, 5.8 Hz), 147.4 (ddt, J = 250.8, 14.6, 3.7 Hz), 160.3, 161.8; ^{19}F NMR (acetone-d_6, C_6F_6): δ −136.98 (dd, J = 20.4, 12.0 Hz, 2F), −140.31 (dd, J = 20.4, 12.0 Hz, 2F); IR (KBr): ν 3680, 2948, 2211, 1704, 1601, 1479, 1330, 1254, 1171, 995, 836 cm^{-1}; HRMS (FAB): [M+] calcd $C_{22}H_{20}F_4O_3$: 408.1349, found: 408.1343.

2.2.8. 2,3,5,6-Tetrafluoro-4-{2-(4-octyloxy)ethyn-1-yl}benzoic acid (**1h**)

Yield: 41% (white solid); T_m: 167 °C (determined by POM); ^{1}H NMR (DMSO-d_6): δ 0.86 (t, J = 6.8 Hz, 3H), 1.23–1.34 (m, 8H), 1.40 (quin, J = 6.8 Hz, 2H), 1.72 (quin, J = 6.8 Hz, 2H), 4.02 (t, J = 6.8 Hz, 2H), 7.02 (d, J = 8.8 Hz, 2H), 7.56 (d, J = 8.8 Hz, 2H), 14.6 (brs, 1H); ^{13}C NMR (DMSO-d_6): δ 13.9, 22.0, 25.4, 24.5, 28.6, 28.7, 31.2, 67.8, 72.8 (t, J = 4.4 Hz), 103.6 (t, J = 3.6 Hz), 106.0 (t, J = 16.9 Hz), 111.8, 113.6 (t, J = 17.6 Hz), 115.2, 133.6, 142.3–145.2 (m, 1C), 144.5–147.4 (m, 1C), 160.0, 160.3; ^{19}F NMR (DMSO-d_6, CFCl$_3$): δ −136.49 (dd, J = 23.3, 10.9 Hz, 2F), −140.50 (dd, J = 23.3, 10.9 Hz, 2F); IR (KBr): ν 3673, 2946, 2210, 1704, 1600, 1477, 1329, 1253, 1170, 995, 836 cm^{-1}; HRMS (FAB): [M+] calcd C$_{23}$H$_{22}$F$_4$O$_3$: 422.1505, found: 422.1510.

2.3. Single-Crystal X-ray Diffraction

Single-crystal X-ray diffraction (XRD) spectra were recorded using an XtaLAB AFC11 diffractometer (Rigaku, Tokyo, Japan). The reflection data were integrated, scaled, and averaged using the CrysAlisPro program (ver. 1.171.39.43a; Rigaku Corporation, Akishima, Japan), while empirical absorption corrections were applied using the SCALE 3 AB-SPACK scaling algorithm (CrysAlisPro). The structures were identified by a direct method (SHELXT-2018/2 [19]) and refined using the full matrix least-squares method (SHELXL-2018/3 [20]) visualized by Olex2 [21]. Crystallographic data were deposited in the Cambridge Crystallographic Data Centre (CCDC) database (CCDC 2193549 for **1a** and 2193550 for **1e**), which were obtained free of charge from the CCDC at www.ccdc.cam.ac.uk/data_request/cif (accessed on 23 November 2022).

2.4. Phase Transition Behavior

The phase transition behaviors were observed by POM using an Olympus BX53 microscope (Tokyo, Japan) equipped with a cooling and heating stage (10002L, Linkam Scientific Instruments, Surrey, UK). Thermodynamic characterization was performed by differential scanning calorimetry (DSC; DSC-60 Plus, Shimadzu, Kyoto, Japan) at heating and cooling rates of 5.0 °C min^{-1} under N$_2$.

2.5. Photophysical Properties

Ultraviolet–visible (UV–vis) absorption spectra were recorded using a JASCO V-750 absorption spectrometer (JASCO, Tokyo, Japan). The PL spectra of the solutions were measured using an FP-6600 fluorescence spectrometer (JASCO, Tokyo, Japan). The PL quantum yields were measured using a Quantaurus-QY C11347-01 instrument (Hamamatsu Photonics, Hamamatsu, Japan).

2.6. Theoretical Calculations

All computations were performed using Gaussian 16 program set [22] with the density functional theory (DFT) at the M06-2X hybrid functional [23] and 6-31+G(d,p) (for all atoms) basis set with a conductor-like polarizable continuum model (CPCM) [24] for CH$_2$Cl$_2$. Theoretical vertical transitions were also calculated using the time-dependent DFT (TD-DFT) method at the same theory level using the same solvation model.

3. Results and Discussion

3.1. Synthesis and Crystal Structure

We first synthesized the fluorinated tolanecarboxylic acid **1** from the corresponding ester **2** via hydrolysis under basic conditions; synthesis of **2** was previously accomplished (Figure 3) [12].

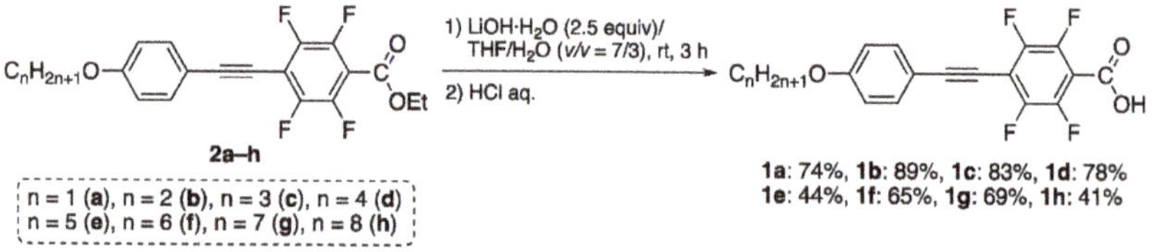

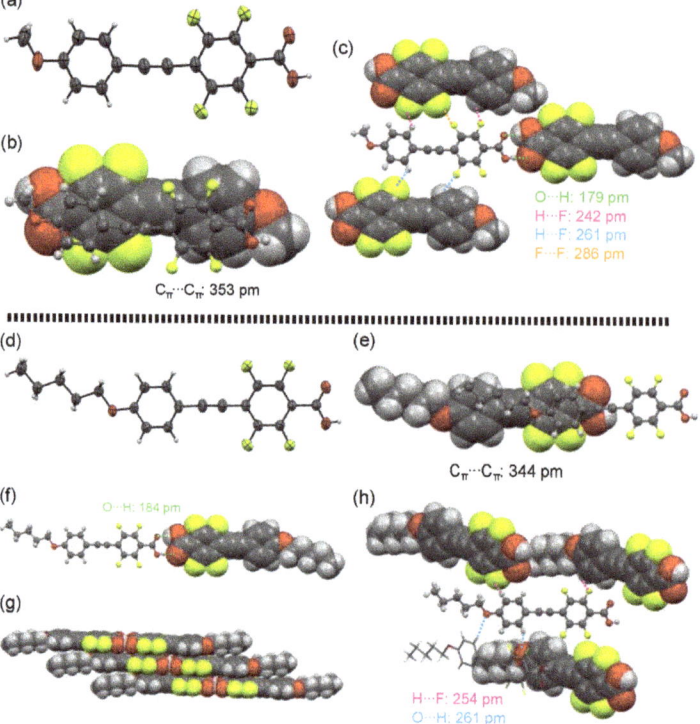

Figure 3. Synthetic procedure of **1a–h** from the corresponding ester **2**.

Treatment of ester **2a** with 2.5 equivalent of LiOH·H$_2$O in a mixed solvent of THF and H$_2$O (v/v = 7/3) at room temperature for 3 h underwent a hydrolysis reaction, which proceeded smoothly. Subsequently, treatment with an aqueous solution of concentrated HCl produced corresponding fluorinated tolanecarboxylic acid **1a**. The product was purified by column chromatography and recrystallization, and the resulting **1a** was generated as a white solid in a 74% isolated yield. Using a similar synthetic procedure, other analogs **1b–h** bearing various alkoxy chains were also produced in 41–89% isolated yields. The molecular structures of **1a–h** were assessed by ^{1}H, ^{13}C, and ^{19}F-NMR, along with IR and HRMS. All structures were completely identified and sufficiently pure to evaluate their phase transition and photophysical behaviors.

Among the fluorinated carboxylic acids **1a–h**, methoxy-substituted **1a** and pentyloxy-substituted **1e** afforded single crystals that were appropriate for X-ray crystallographic analysis. Figure 4 shows the crystal structures of **1a** and **1e** and their packing structures.

Figure 4. (**a**) Crystal structure of **1a** with an ORTEP drawing and (**b**,**c**) packing structures. (**d**) Crystal structure of **1e** with an ORTEP drawing and (**e–h**) packing structures.

Methoxy-substituted **1a** crystalized with a triclinic system in the *P*–1 space group and two molecular units were contained in the Cry lattice. The dihedral angle between two aromatic rings in the tolane scaffold was only 4.7°, almost coplanar to each other (Figure 4a). The dihedral angle between the fluorinated aromatic ring and the carbonyl plane was reported 34° for the ester precursor **2a** [12]. However, the dihedral angle of the carboxylic acid **1a** was only 3.7°, resulting in an almost coplanar structure. With respect to the packing structures, the two planar tolane scaffolds were arranged in a layer structure with an antiparallel direction. This phenomenon is caused by the electrostatic weak π–π interaction (short contact of Cπ···Cπ: 353 pm) between the electron-rich methoxy-substituted aromatic ring and the electron-deficient fluorinated aromatic ring (Figure 4b). Additionally, the fluorinated tolanecarboxylic acid **1a** formed plural intermolecular interactions (Figure 4c), such as O···H hydrogen bond (short contact of O···H: 179 pm), H···F hydrogen bond (short contact of H···F: 242 and 261 pm), and F···F interaction (short contact of F···F: 286 pm), wherein the short contacts mentioned above were almost identical or below the sum of van der Waals radii [25].

In contrast, pentyloxy-substituted **1e** furnished single crystals with a monoclinic system in the *C* 2/*c* space group, and eight molecular units were contained in the Cry lattice. The electron-rich aromatic ring and the electron-deficient fluorinated aromatic ring were nearly coplanar, with a deviation of 3.0°. The dihedral angle between the fluorinated aromatic ring and the carbonyl plane was 11°, being almost coplanar (Figure 4d). However, unlike the π–π stacking of the antiparallel orientation in **1a**, **1e** formed a slipped π–π stacking (short contact of Cπ···Cπ: 344 pm) with a synparallel orientation induced by the electrostatic interaction between the electron-rich pentyloxy aromatic ring and the electron-deficient fluorinated aromatic ring (Figure 4e). As shown in Figure 4f,g, the carboxyl units in **1e** also formed an intermolecular O···H hydrogen bond with a short contact of 184 pm, leading to formation of layer structures. For construction of the crystal structure of **1e**, more intermolecular interactions, such as additional O···H and H···F hydrogen bonds (Figure 4h), were also observed. The interatomic distance was 254 and 261 pm for O···H and H···F, respectively, which was also almost identical or below the sum of van der Waals radii [25].

3.2. Phase Transition Behavior

With the fluorinated tolanecarboxylic acids, **1a–h**, in hand, we evaluated their phase transition behavior using DSC and POM. Table 1 summarizes the phase sequence and phase transition temperature for **1a–h** during the first heating and cooling process determined by POM. Subsequent phase transition behavior is listed in Table S2 (ESI). Figure 5 shows the POM texture images of **1d–h** observed in the mesophase.

Table 1. The phase transition behavior of the fluorinated tolanecarboxylic acids, **1a–h**, during the first heating and cooling process observed by POM.

Molecule	Phase Sequence and Phase Transition Temperature [°C] [1]	
	Heating Process	Cooling Process
1a	Cry 224 Iso	Iso 165 G
1b	Cry 207 Iso	Iso 140 G
1c	Cry 211 Iso	Iso 158 G
1d	Cry 176 N 198 Iso	Iso 124 G
1e	Cry [1] 160 Cry [2] 170 N 191 Iso	Iso 132 G
1f	Cry 181 N 190 Iso	Iso 141 G
1g	Cry 169 N 184 Iso	Iso 150 G
1h	Cry 161 N 186 Iso	Iso 146 N 118 G

[1] Determined by POM. Abbreviations: Cry: crystalline; G: Glassy; N: nematic; and Iso: isotropic phases.

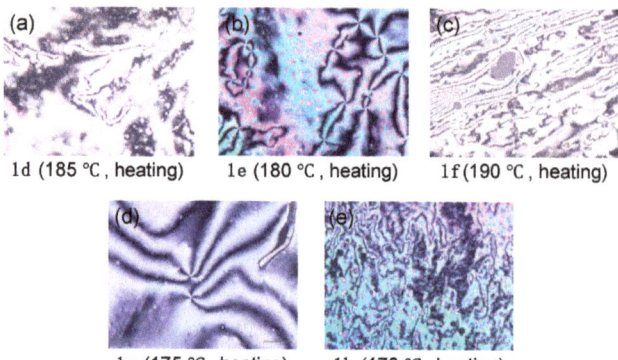

Figure 5. Optical microphotographs of (**a**) **1d**, (**b**) **1e**, (**c**) **1f**, (**d**) **1g**, and (**e**) **1h** in the mesophase phase.

The DSC measurement of methoxy-substituted **1a** showed a large endothermic peak with an onset temperature of 223 °C during the first heating process. No sharp exothermic peak due to the Iso → Cry phase transition was observed during the subsequent cooling process. As a result of POM observation, however, a phase transition from the Iso phase to a glassy amorphous solid (G) phase was observed; **1a** did not show any mesophase (Figure S25). The POM observation also proved that **1a** showed no mesophase between the Cry and isotropic (Iso) phases. Additionally, no mesophase was observed for ethoxy-substituted **1b** and propoxy-substituted **1c** by POM and DSC measurements. In contrast, butoxy-substituted **1d** showed an endothermic phase transition between the Cry and Iso phases in the first heating process of the DSC measurement and a bright-viewing field with fluidity during the heating and cooling processes of the POM observation. Thus, the phase transition behavior of butoxy-substituted **1d** possessed the LC characteristic. A four-brushed Schlieren texture was observed as the optical image (Figure 5a), which is a typical texture for the N LC phase. During the subsequent cooling process, however, only the Iso → G phase transition was observed. The phase transition behavior was also supported by temperature-varying powder X-ray diffraction (VT-PXRD) measurements (Figure S26). Further POM observation was found to show similar phase transition between G and Iso phases during the second cycles. Similar to **1d**, molecules **1e** and **1f** also exhibited an N-phase during the first heating process (Figure 5b,c), while, after the first cooling process, the LC phase disappeared, showing only a phase transition between the G and Iso phases. The other analogs, viz., **1g** and **1h**, with a relatively long alkoxy chain, were found to show a mesophase during both heating and cooling processes due to increasing stabilization of the mesophase. Thus, both **1g** with a C_7H_{15} chain and **1h** with a C_8H_{15} chain exhibited an N-phase with a four-brush Schlieren texture through POM measurements during both heating and cooling cycles (Figure 5d,e), in which the observed mesophase can be assigned as an N-phase by the VT-PXRD measurements (Figure S26). Focusing on the melting temperature (T_m), which is defined as the phase transition temperature from Cry to Iso or LC phases, the T_m of **1a–h** was in the range of 167–224 °C for the heating process, which was much higher than that of the corresponding ester derivatives **2a–h** (34–109 °C) [12]. Unlike the ester derivatives, the carboxylic acids exhibited LC phases even with relatively short alkoxy chains, particularly C_4H_9O, due to the increased aspect ratio of the mesogenic core induced by formation of a dimeric structure through hydrogen bonds.

3.3. Photophysical Behavior in Solution Phase

A solution sample was prepared to investigate the photophysical behavior of the fluorinated carboxylic acids, **1a–h**, in the solution phase by individually dissolving **1a–h** in CH_2Cl_2; the concentration was adjusted to 1.0×10^{-5} mol L^{-1}. Figure 6 illustrates the photophysical behavior in the solution, and the photophysical data are summarized in Table 2.

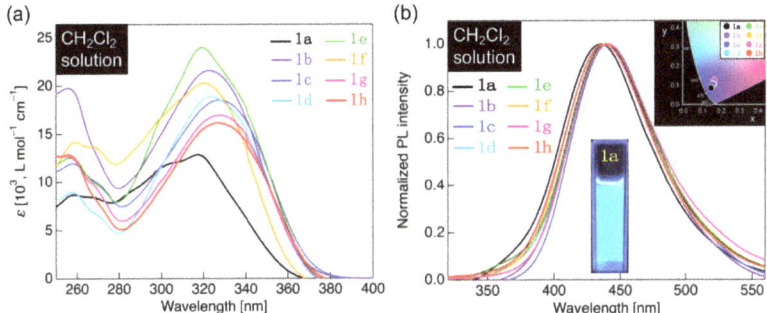

Figure 6. (a) Ultraviolet (UV)–visible absorption spectrum of **1a–h** in the CH$_2$Cl$_2$ solution (concentration: 1.0×10^{-5} mol L^{-1}). (b) PL spectrum of **1a–h** in the CH$_2$Cl$_2$ solution (concentration: 1.0×10^{-5} mol L^{-1}) and a photograph of the PL behavior of **1a** solution under UV light (λ_{ex} = 365 nm). Inset: Commission Internationale de l'Eclairage (CIE) diagram for PL color of **1a–h** solutions.

Table 2. Photophysical data of **1a–h** in solution state.

Molecule	Solvent (E$_T$30)	λ_{abs} [nm] [1] (ε, 10^3 [L mol^{-1} cm^{-1}])	λ_{PL} [nm] [1] (Φ_{PL}) [2]	CIE Coordinate (x, y)
1a	CH$_2$Cl$_2$ (40.7)	259 (8.69), 317 (12.9)	435 (0.33)	(0.150, 0.085)
	Toluene (33.9)	324 (27.3)	403 (0.35)	(0.160, 0.029)
	CHCl$_3$ (39.1)	274 (36.0), 284 (36.2) 317 (27.7)	420 (0.33)	(0.155, 0.051)
	MeCN (45.6)	256 (14.2), 299sh (26.3), 314 (30.2)	463 (0.10)	(0.194, 0.238)
1b	CH$_2$Cl$_2$ (40.7)	256 (19.8), 323 (21.6)	439 (0.33)	(0.152, 0.088)
1c	CH$_2$Cl$_2$ (40.7)	258 (12.0), 327 (18.5)	441 (0.35)	(0.158, 0.103)
1d	CH$_2$Cl$_2$ (40.7)	258 (8.98), 323 (18.9)	439 (0.37)	(0.161, 0.108)
1e	CH$_2$Cl$_2$ (40.7)	257 (12.6), 319 (24.0)	440 (0.27)	(0.160, 0.107)
1f	CH$_2$Cl$_2$ (40.7)	259 (14.2), 320 (20.3)	437 (0.33)	(0.160, 0.101)
1g	CH$_2$Cl$_2$ (40.7)	255 (12.8), 329 (16.9)	440 (0.38)	(0.167, 0.125)
1h	CH$_2$Cl$_2$ (40.7)	256 (12.9), 327 (16.2)	438 (0.37)	(0.163, 0.105)

[1] Concentration: 1.0×10^{-5} mol L^{-1}. [2] Measured using an integrating sphere.

Methoxy-substituted **1a** in CH$_2$Cl$_2$ absorbed UV light with a maximum absorption wavelength (λ_{abs}) near 259 nm and 317 nm (Figure 6a). Other analogs, particularly **1b–h**, also showed two absorption bands: a high-energy absorption band near 255–259 nm of λ_{abs} and a low-energy absorption band near 319–329 nm of λ_{abs} (Figure 6a). Quantum chemical calculations were performed by the TD-DFT method using **1a** as a representative, and two allowed transitions with theoretical absorption wavelengths of 319 and 262 nm were calculated as theoretical vertical transitions (Figure S31). The calculated absorption wavelengths were close to the experimentally obtained λ_{abs}. Thus, the result confirms that the long-wavelength absorption band of **1a** in CH$_2$Cl$_2$ is the ππ* transition with an intramolecular charge transfer (ICT) character involving the highest occupied molecular orbital to the lowest unoccupied molecular orbital (HOMO → LUMO) transition, while the short-wavelength band is the ππ* transition with a local excitation character involving a HOMO−1 → LUMO transition.

With the λ_{abs} as the excitation wavelength, the methoxy-substituted **1a** in the solution state was observed to emit blue PL, with a maximum PL wavelength (λ_{PL}) of approximately 435 nm (Figure 6b). In addition, **1b–h** with varying lengths of alkoxy group were found to have a PL band with λ_{PL} in the range of 437–441 nm, leading to the blue PL. Considering the observed PL colors using the Commission Internationale de l'Eclailage (CIE) diagrams

(Figure 6b, inset), the CIE coordinates for the PL colors of **1a–h** were similar to each other. The PL color of the fluorinated tolanecarboxylic acids in CH$_2$Cl$_2$ showed a uniform blue PL in the solution state without affecting the alkoxy-chain length. PL quantum yields (Φ_{PL}) of **1a–h** in CH$_2$Cl$_2$ solutions were in the range of 0.27–0.38, which is higher than that of the unsubstituted tolane [26,27]. This phenomenon is observed because the donor–π–acceptor structure of the fluorinated tolanecarboxylic acid suppresses the internal conversion from the ππ* excited state to the dark πσ* excited state. In addition, we investigated the effect of solvent polarity on photophysical properties using **1a** as a representative [28]. We found that, although the solvent polarity did not affect the absorption properties significantly, the PL properties shifted to longer wavelengths as the polarity increased, which is attributed to stabilization of solute–solvent interactions (Figure S28d). Considering the Lippert–Mataga plot [29,30], which is created from the orientational polarizability (Δf) and Stokes shift ($\Delta \nu$) on the horizontal and vertical axes, respectively, a linear relationship was obtained (Figure S28e). The dipole moment difference ($\mu_e - \mu_g$) between the excited and ground states was 14.1 D, which was calculated from the slope of the straight line. The large difference in the dipole moment proves that the radiative deactivation from the ICT excited state resulted in the PL of **1a–h**.

3.4. Photophysical Behavior in Aggregated Phases

We next examined the PL behavior of fluorinated tolanecarboxylic acids, **1a–h**, in the aggregated phases. Figure 7 shows the PL spectrum, photographs of the PL behavior under UV irradiation, and a CIE diagram for the PL colors. The photophysical data of **1a–h** in the aggregated phases are summarized in Table 3.

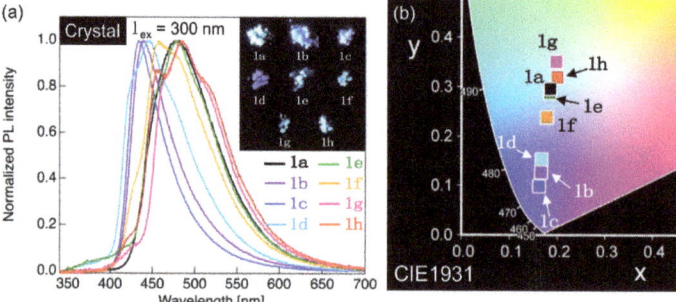

Figure 7. (a) PL spectra of **1a–h** in crystalline state. Excitation wavelength (λ_{ex}): 300 nm. Inset: Photographs of the PL behavior of the **1a–h** crystals under UV light (λ_{ex} = 365 nm). (b) CIE color diagram of PL colors for **1a–h** crystals.

Table 3. Photophysical data of **1a–h** in aggregated phases.

Molecule	Phase [1]	λ_{PL} [nm] (Φ_{PL}) [2]	CIE Coordinate (x, y)
1a	Crystalline (Cry)	481 (0.99)	(0.185, 0.296)
1b	Cry	440 (0.69)	(0.166, 0.127)
1c	Cry	434 (0.71)	(0.162, 0.099)
1d	Cry	428sh, 446(0.63)	(0.167, 0.153)
1e	Cry	454sh, 478 (0.57)	(0.184, 0.288)
1f	Cry	430sh, 458, 480sh (0.60)	(0.178, 0.239)
1g	Cry	465sh, 482, 511sh (0.49)	(0.198, 0.351)
1h	Cry	455, 483, 511sh (0.70)	(0.200, 0.320)
1h	–[3]	428sh, 454, 479 (0.12)	(0.189, 0.240)

[1] Unless mentioned otherwise, the crystalline sample prepared by column chromatography and recrystallization was used. Measured at 25 °C. [2] Measured using an integrating sphere. [3] Samples with mesophase aggregate structures were prepared by quenching and immersing mesophase (170 °C) during the 1st heating process in liquid nitrogen to maintain the mesophase molecular aggregates at room temperature.

When methoxy-substituted **1a** in the Cry phase was excited by irradiation with incident light of 300 nm, which is the maximum excitation wavelength (λ_{ex}), a single PL band was observed with a λ_{PL} of approximately 481 nm (Figure 7a). As shown in Figure 8b,c, the CIE coordinate (x, y) of the PL color was (0.185, 0.296), indicating that the PL color was light blue. Notably, a CH_2Cl_2 solution of **1a** had a Φ_{PL} of 0.33, whereas the **1a** in the Cry phase dramatically increased the Φ_{PL} to up to 0.99. Although **1b–h** with varying lengths of the alkoxy chain had almost identical λ_{PL} in dilute solutions, they exhibited various λ_{PL} in the Cry phase, ranging from 428 to 511 nm (Figure 7a). The alteration in λ_{PL} indicated a change in the PL color. Thus, various PL colors ranging from blue to light green were obtained by changing the length of the alkoxy chain, which is evident from the photographs and the CIE diagram demonstrating the PL colors (Figure 7b). The Φ_{PL} values of **1b–h** in the Cry phase were in the range of 0.49–0.71, which were higher than those in dilute solutions (up to 0.38). Considering the crystal packing structures of **1a** and **1e** shown in Figure 4, the change in the length of the alkoxy chain considerably altered the molecular arrangements in molecular aggregated phases; **1a–h** exhibited various PL behaviors in the Cry phase. Furthermore, O···H and H···F hydrogen bonds and intermolecular interactions, such as F···F interactions and weak π···π interactions, function in the Cry phase, which possibly restricts the molecular motion to suppress non-radiative deactivation, resulting in a significant increase in the Φ_{PL} in the Cry state.

The PL behavior in the aggregated structures of the N-phase was evaluated using octyloxy-substituted **1h** with an N LC phase. The measurement sample was prepared by quenching the sample with the N LC phase at 170 °C, which was developed during the 1st heating process, with liquid nitrogen. Figure 8 shows the PL spectra and CIE diagrams for **1h** with the Cry- and N-phase molecular aggregated structures.

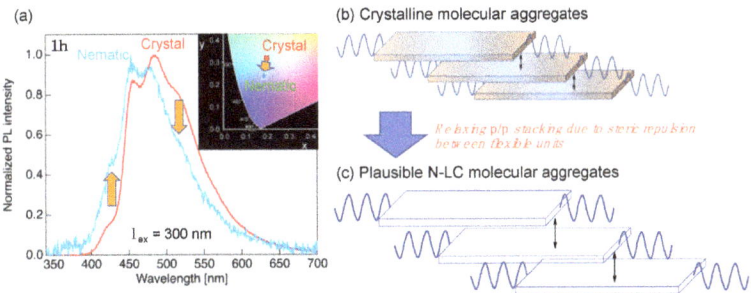

Figure 8. (**a**) PL spectra of **1h** with the Cry- and N-phase molecular aggregated structures. Inset: CIE diagram of the PL colors for **1h** with the Cry- and N-phase aggregated structures. (**b**,**c**) Schematic illustration of plausible structural alteration from the Cry- to N-phase molecular aggregated structures.

The PL spectrum of **1h** with N-phase aggregated structures was also obtained by irradiation with incident light of 300 nm, in which the λ_{PL} was approximately 454 and 479 nm, along with a shoulder peak of approximately 428 nm. Compared to the Cry phase, the PL spectra of the N-phase aggregated structure yielded a slight short-wavelength shift with weakened long-wavelength shoulder peaks and increased short-wavelength peaks. In the Cry phase, the dimer mesogens formed a dense packing structure due to the weak π···π interactions (Figure 8b), as shown in Figure 4g. Conversely, the phase transition to the N-phase increased the molecular fluidity, allowing the increase in the two molecular distances (Figure 8c). The increased spacing between the dimer mesogens in the N-phase aggregated structure and the promotion of the molecular motion drastically reduced the Φ_{PL} compared to that in the Cry phase.

4. Conclusions

In conclusion, we designed and synthesized a series of fluorinated tolanecarboxylic acids bearing various lengths of alkoxy chains and investigated their phase transition and photophysical behaviors. The fluorinated tolanecarboxylic acids exhibited the N LC phase due to formation of the dimer-type mesogen of the carboxylic acid moiety via O···H hydrogen bonds. Furthermore, regarding photophysical measurements, the fluorinated tolanecarboxylic acids emitted blue PL in the solution phase. The PL quantum yield (Φ_{PL}) was approximately 0.33, which was higher than that of the unsubstituted tolane. The fluorinated tolanecarboxylic acid exhibited remarkably strong PL even in the Cry phase, and its Φ_{PL} was much higher than that in the dilute-solution state, which could be attributed to the O···H and H···F hydrogen bonds and the weak π···π and F···F intermolecular interactions. Investigation of the PL behavior in the N-phase molecular aggregated structure revealed a slight short-wavelength shift and a significant decrease in Φ_{PL}, which is attributable to the wider spacing between the dimer-type mesogens caused by increasing the molecular fluidity in the N-phase. These findings will offer a new molecular design for PLLC molecules effectively using intermolecular interactions and pave the way for developing new thermo-responsive luminescent materials in the future.

Supplementary Materials: The following supporting information can be downloaded at: https://www.mdpi.com/article/10.3390/cryst13010025/s1, Figures S1–S24: NMR spectra of **1a–h**; Figure S25: DSC thermograms of **1a–h**; Figure S26: TG thermograms of **1a–h**; Figure S27: PXRD patterns of **1d–h** on the mesophase; Figure S28: UV–vis absorption and PL spectra of **1a–h** in CH_2Cl_2 solution; Figure S29: UV–vis absorption and PL spectra of **1a** in different solvents; Figure S30: Excitation and PL spectra of **1a–h** in the Cry phase; Figure S31: Excitation and PL spectra of **1h** in the aggregated structure of the nematic phase; Figures S32 and S33: Optimized structure of **1a** and **1e** and their orbital distributions; Table S1: Crystallographic data of **1a** and **1e**; Table S2: Phase transition behaviors of **1a–h** observed by DSC measurements; Table S3: Solvent effect on the photophysical behavior; Tables S4 and S5: Cartesian coordinate for **1a** and **1e**.

Author Contributions: Conceptualization, S.Y.; methodology, S.Y.; validation, S.Y., M.K., K.Y. and T.K.; investigation, S.Y., M.K. and K.Y.; resources, S.Y. and T.K.; data curation, S.Y.; writing—original draft preparation, S.Y. and M.K.; writing—review and editing, S.Y., M.K., K.Y., M.N., T.A., H.F. and T.K.; visualization, S.Y.; supervision, S.Y.; project administration, S.Y.; funding acquisition, S.Y. All authors have read and agreed to the published version of the manuscript.

Funding: This research was partially funded by the Murata Science Foundation and Shorai Foundation for Science and Technology.

Data Availability Statement: Not applicable.

Acknowledgments: We are deeply grateful to Sakurai and Shimizu (Kyoto Inst. Tech.) for their valuable cooperation in the PXRD measurements.

Conflicts of Interest: The authors declare no conflict of interest.

References

1. Irfan, M.; Sumra, I.; Zhang, M.; Song, Z.; Liu, T.; Zeng, Z. Thermochromic and highly tunable color emitting bis-tolane based liquid crystal materials for temperature sensing devices. *Dye. Pigment.* **2021**, *190*, 109272. [CrossRef]
2. Kato, T.; Uchida, J.; Ichikawa, T.; Sakamoto, T. Functional liquid crystals towards the next generation of materials. *Angew. Chem. Int. Ed.* **2018**, *57*, 4355–4371. [CrossRef] [PubMed]
3. Wang, Y.; Shi, J.; Chen, J.; Zhu, W.; Baranoff, E. Recent progress in luminescent liquid crystal materials: Design, properties and application for linearly polarized emission. *J. Mater. Chem. C* **2015**, *3*, 7993–8005. [CrossRef]
4. Sagara, Y.; Kato, T. Stimuli-responsive luminescent liquid crystals: Change of photoluminescent colors triggered by a shear-induced phase transition. *Angew. Chem. Int. Ed.* **2008**, *120*, 5253–5256. [CrossRef]
5. Zhao, D.; Fan, F.; Cheng, J.; Zhang, Y.; Wong, K.S.; Chigrinov, V.G.; Kwok, H.S.; Guo, L.; Tang, B.Z. Light-emitting liquid crystal displays based on an aggregation-induced emission luminogen. *Adv. Opt. Mater.* **2015**, *3*, 199–202. [CrossRef]
6. Yamada, S.; Miyano, K.; Konno, T.; Agou, T.; Kubota, T.; Hosokai, T. Fluorine-containing bistolanes as light-emitting liquid crystalline molecules. *Org. Biomol. Chem.* **2017**, *15*, 5949–5958. [CrossRef]

7. Morita, M.; Yamada, S.; Agou, T.; Kubota, T.; Konno, T. Luminescence tuning of fluorinated bistolanes via electronic or aggregated-structure control. *Appl. Sci.* **2019**, *9*, 1905. [CrossRef]
8. Morita, M.; Yamada, S.; Konno, T. Fluorine-induced emission enhancement of tolanes via formation of tight molecular aggregates. *New J. Chem.* **2020**, *44*, 6704–6708. [CrossRef]
9. Morita, M.; Yamada, S.; Konno, T. Systematic studies on the effect of fluorine atoms in fluorinated tolanes on their photophysical properties. *Molecules* **2021**, *26*, 2274. [CrossRef]
10. Yamada, S.; Kobayashi, K.; Morita, M.; Konno, T. D–π–A-type fluorinated tolanes with a diphenylamio group: Crystal polymorphism formation and photophysical behavior. *CrystEngComm* **2022**, *24*, 936–941. [CrossRef]
11. Morita, M.; Yamada, S.; Konno, T. Halogen atom effect of fluorinated tolanes on their luminescence characteristics. *New J. Chem.* **2022**, *46*, 4562–4569. [CrossRef]
12. Yamada, S.; Kataoka, M.; Yoshida, K.; Nagata, M.; Agou, T.; Fukumoto, H.; Konno, T. Photophysical and thermophysical behavior of D-p-A-type fluorinated diphenylacetylenes bearing an alkoxy and an ethoxycarbonyl group at both longitudinal molecular terminals. *J. Fluor. Chem.* **2022**, *261–262*, 110032. [CrossRef]
13. Yamada, S.; Uto, E.; Sakurai, T.; Konno, T. Development of thermoresponsive near-ultraviolet photoluminescent liquid crystals using hexyloxy-terminated fluorinated tolane dimers connected with an alkylene spacer. *J. Mol. Liq.* **2022**, *362*, 119755. [CrossRef]
14. Yamada, S.; Uto, E.; Yoshida, K.; Sakurai, T.; Konno, T. Development of photoluminescent liquid-crystalline dimers bearing two fluorinated tolane-based luminous mesogens. *J. Mol. Liq.* **2022**, *363*, 119884. [CrossRef]
15. Arakawa, Y.; Sasaki, Y.; Igawa, K.; Tsuji, H. Hydrogen bonding liquid crystalline benzoic acids with alkylthio groups: Phase transition behavior and insights into the cybotactic nematic phase. *New J. Chem.* **2017**, *41*, 6514–6522. [CrossRef]
16. Arakawa, Y.; Kang, S.; Watanabe, J.; Konishi, G. Assembly of thioether-containing rod-like liquid crystalline materials assisted by hydrogen-bonding terminal carboxyl groups. *RSC Adv.* **2015**, *5*, 8056–8062. [CrossRef]
17. Wen, J.; Tian, M.; Yu, H.; Guo, Z.; Chen, Q. Novel fluorinated liquid crystals. Part 9.—Synthesis and mesomorphic properties of 4-(n-alkoxycarbonyl)phenyl 4-[(4-n-alkoxy-2,3,5,6-tetrafluorophenyl)ethynyl]benzoates. *J. Mater. Chem.* **1994**, *4*, 327–330. [CrossRef]
18. Wen, J.X.; Tian, M.Q.; Chen, Q. Synthesis and mesomorphic properties of 4'-n-alkoxy-2,3,5,6-tetrafluorobiphenyl-4-carboxylic acids. *J. Fluor. Chem.* **1994**, *67*, 207–210. [CrossRef]
19. Sheldrick, G.M. SHELXT-Integrated space-group and crystal-structure determination. *Acta Crystallogr. Sect. A Found. Adv.* **2015**, *71*, 3–8. [CrossRef]
20. Sheldrick, G.M. Crystal structure refinement with SHELXL. *Acta Crystallogr. Sect. C Struct. Chem.* **2015**, *71*, 3–8. [CrossRef]
21. Dolomanov, O.V.; Bourhis, L.J.; Gildea, R.J.; Howard, J.A.K.; Puschmann, H. OLEX2: A complete structure solution, refinement, and analysis program. *J. Appl. Crystallogr.* **2009**, *42*, 339–341. [CrossRef]
22. Frisch, M.J.; Trucks, G.W.; Schlegel, H.B.; Scuseria, G.E.; Robb, M.A.; Cheeseman, J.R.; Scalmani, G.; Barone, V.; Petersson, G.A.; Nakatsuji, H.; et al. *Gaussian 16, Revision B.01*; Gaussian, Inc.: Wallingford, CT, USA, 2016.
23. Hohenstein, E.G.; Chill, S.T.; Sherrill, C.D. Assessment of the performance of the M05-2X and M06-2X exchange-correlation functionals for noncovalent interactions in biomolecules. *J. Chem. Theory Comput.* **2008**, *4*, 1996–2000. [CrossRef] [PubMed]
24. Li, H.; Jensen, J.H. Improving the efficiency and convergence of geometry optimization with the polarizable continuum model: New energy gradients and molecular surface tesselation. *J. Comput. Chem.* **2004**, *25*, 1449–1462. [CrossRef] [PubMed]
25. Bondi, A. van der Waals volumes and radii. *J. Phys. Chem.* **1964**, *68*, 441–451. [CrossRef]
26. Zgierski, M.Z.; Lim, E.C. Nature of the 'dark' state in diphenylacetylene and related molecules: State switch from the linear $\pi\pi^*$ state to the bent $\pi\sigma^*$ state. *Chem. Phys. Lett.* **2004**, *387*, 352–355. [CrossRef]
27. Saltiel, J.; Kumar, V.K.R. Photophysics of diphenylacetylene: Light from the "dark state". *J. Phys. Chem. A* **2012**, *116*, 10548–10558. [CrossRef]
28. Reichardt, C. Solvatochromic dyes as solvent polarity indicators. *Chem. Rev.* **1994**, *94*, 2319–2358. [CrossRef]
29. Mataga, N.; Kaifu, Y.; Koizumi, M. The solvent effect on fluorescence spectrum, change of solute-solvent interaction during the lifetime of excited solute molecule. *Bull. Chem. Soc. Jpn.* **1955**, *28*, 690–691. [CrossRef]
30. Mataga, N.; Kaifu, Y.; Koizumi, M. Solvent effects upon fluorescence spectra and dipole moments of excited molecules. *Bull. Chem. Soc. Jpn.* **1956**, *29*, 465–470. [CrossRef]

Disclaimer/Publisher's Note: The statements, opinions and data contained in all publications are solely those of the individual author(s) and contributor(s) and not of MDPI and/or the editor(s). MDPI and/or the editor(s) disclaim responsibility for any injury to people or property resulting from any ideas, methods, instructions or products referred to in the content.

Article

Photoalignment and Photofixation of Chromonic Mesophase in Ionic Linear Polysiloxanes Using a Dual Irradiation System

Mitsuo Hara [1,*], Ayaka Masuda [1], Shusaku Nagano [2] and Takahiro Seki [1,*]

[1] Department of Molecular and Macromolecular Chemistry, Graduate School of Engineering, Nagoya University, Furo-cho, Chikusa-ku, Nagoya 464-8603, Aichi, Japan

[2] Department of Chemistry, College of Science, Rikkyo University, 3-34-1 Nishi-Ikebukuro, Toshima, Tokyo 171-8501, Japan

* Correspondence: mhara@chembio.nagoya-u.ac.jp (M.H.); tseki@chembio.nagoya-u.ac.jp (T.S.)

Abstract: Photoalignment technology enables macroscopic alignment of liquid crystalline molecules and their aggregates in a non-contact process by irradiating photo-responsive liquid crystalline compounds with linearly polarized light. Because photoalignment techniques prevent dust generation and uneven stretching, and accomplish fine and complex patterning, they are involved in the practical process of fabricating display panels, and continue to be applied in the research and creation of various anisotropic materials. Brilliant yellow (BY), a chromonic liquid crystal, has attracted considerable attention as the photoalignment sublayer in recent years, because of its ability to induce a high dichroic nature among many photo-responsive liquid crystalline materials. However, its dichroism is not maintained after prolonged exposure to a humid environment because of its intrinsic strong hygroscopicity of ionic BY molecules. In this study, to overcome this drawback, the photoalignment and successive photo-fixation of the BY columnar phase is proposed using UV-curable ionic polysiloxane as a matrix. Visible light was used for the photoalignment of the BY columnar phase, and UV light for photo-fixation. Consequently, the columnar chromonic phase is found to retain its orientation even after 4 h of exposure to a highly humid environment.

Keywords: chromonic liquid crystal; polysiloxane; photoalignment; UV curing

Citation: Hara, M.; Masuda, A.; Nagano, S.; Seki, T. Photoalignment and Photofixation of Chromonic Mesophase in Ionic Linear Polysiloxanes Using a Dual Irradiation System. *Crystals* **2023**, *13*, 326. https://doi.org/10.3390/cryst13020326

Academic Editor: Borislav Angelov

Received: 4 February 2023
Revised: 10 February 2023
Accepted: 12 February 2023
Published: 15 February 2023

Copyright: © 2023 by the authors. Licensee MDPI, Basel, Switzerland. This article is an open access article distributed under the terms and conditions of the Creative Commons Attribution (CC BY) license (https://creativecommons.org/licenses/by/4.0/).

1. Introduction

Liquid crystals (LCs) self-assemble in response to temperature, electric fields, concentration, light irradiation, and other environmental conditions to form nano-periodic structures. Such soft materials are suitable for templates of nanostructures [1]. The dynamic cooperativity of LCs also facilitates control of the arrangement of molecules and their aggregates over a macroscopic scale, and enables their use in a variety of applications such as biosensors, optical devices, separators, and actuators [2–5]. Among them, lyotropic LCs mostly self-assemble in aqueous solvents, and they can be applied to various environmentally friendly processes. Some lyotropic LCs are classified as chromonic LCs [6–11]. Chromonic LCs are soft materials containing dye groups such as mesogens, which self-assemble to form columnar nanostructures via $\pi\pi$ interaction in solvents. The self-assembly structures that involve absorption anisotropy of chromonic LCs can be applied to optically functional films when the structures are align overlarge areas [12–15]. In particular, the optical manipulation of chromonic LC phases using photoalignment techniques [16–18] has the advantage of ready achievement of fine patterning, which is difficult to be accomplish by conventional mechanical rubbing or film stretching techniques [19–22].

Among chromonic LCs, brilliant yellow (BY) (as shown in Scheme 1) has attracted significant attention in recent years, and research using BY films as LC alignment sublayers has been extensively undertaken [23–38]. This is because the columnar LC phase of BY is well photoaligned using linearly polarized UV or visible light, creating highly dichroic optical films [39–42]. Such photoaligned BY films function well as alignment sublayers for low-molecular-mass nematic LCs. However, BY exhibits a hygroscopic nature, which means that it cannot maintain dichroism for a long time in a highly humid environment. If the photoaligned BY columnar phase can be stabilized, it is anticipated that the high dichroic properties of BY could be used in more applications in wider variety situations; however, such a methodology has not yet been proposed.

Brilliant Yellow (BY)

Scheme 1. Chemical structure of chromonic liquid crystal used in this study.

Matrix fixation by sol-gel reaction is often used in the stabilization of lyotropic LC phases [43,44]. However, chromonic LCs generally have low solubility in metallic-alkoxide sol (such as silica sol), and the addition of a compatibilizer is necessary to improve compatibility [14,45]. In addition, the chromonic LCs are solidified by the silica matrix through film formation, which makes photoalignment of BY after film formation difficult to achieve using conventional stabilization methods for lyotropic LCs.

Some recently developed ionic linear polysiloxanes are compatible with the hydrophilic region of lyotropic LC phases [46]. By introducing functional groups that can be cross-linked by ultraviolet light into polysiloxane, it is also possible to fix the LC phase in the matrix at targeted times and locations after film formation [47]. Conversion of the ionic groups of polysiloxanes is relatively easy, and a variety of ionic polysiloxanes can be prepared [48]. The ease of design of the ionic group of polysiloxanes means that development of an ionic polysiloxanes with high compatibility with chromonic LCs can be easy. In this study, the design for an ionic linear polysiloxane that is compatible with BY and has photo-crosslinking groups is proposed. By preparing a mixed thin film of the polysiloxane and BY, the photo-orientation of the BY columnar phase by visible light is demonstrated in the polysiloxane matrix. Subsequent fixation by UV light irradiation is also achieved. By photo-fixing the BY columnar phase, the dichroism ratio may be maintained even after extended exposure to a humid environment.

2. Materials and Methods

2.1. Materials

Scheme 2 shows the chemical structures used in this paper. BY chromonic dye and Irgacure® 2959 (I2959) photoinitiator were purchased from Sigma-Aldrich and Tokyo Chemical Industry (TCI), respectively. An anionic linear polysiloxane containing vinyl groups (PSSV) was synthesized via polycondensation of silane coupling agents. N, N-Dimethylformamide (DMF) was purchased from Kanto Chemical. The I2959 and DMF used were of commercial purity. Water was obtained through a Direct-Q® 3UV purification system (Millipore Corp., Burlington, MA, USA, ρ (resistivity) > 18 MΩ·cm at 25 °C).

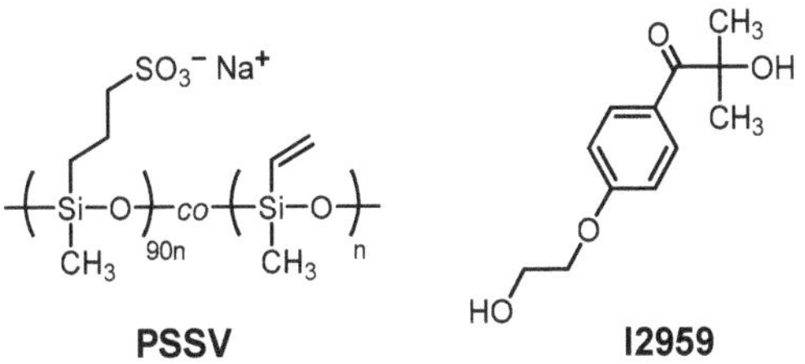

Scheme 2. Chemical structures of hygroscopic siloxane copolymer and photoinitiator used in this study.

2.2. Synthesis of PSSV

PSSV was synthesized according to the scheme shown in Figure 3a [49]. The detailed procedure is follows.

1.8 g (1.0×10^{-2} mol) of 3-mercaptopropyl(dimethoxy)methylsilane (TCI) and 3.5×10^{-2} g (2.6×10^{-4} mol) of dimethoxymethylvinylsilane (TCI) were added to 54 g of 2 mol L^{-1} of sodium hydroxide (TCI). The mixture solution was stirred at 25 °C for 2 h before the addition of polysiloxane containing mercapto groups and vinyl groups. To oxidize the mercapto groups, 6 g of 30% hydrogen peroxide water (Kanto Chemical) was added to the mixture solution, and the resulting solution was stirred at 25 °C for 12 h. The solvent was dried using a smart evaporator C1 (BioChromato) and a white powder was obtained. The powder was dissolved in 300 mL of water, and ion exchange occurred using an ion-exchange resin Amberlite® IR120, in hydrogen form (Fluka). After the solution was treated using an ion-exchange resin IR120B Na (Organo) and dried using an evaporator, 2.1 g of white powder was obtained.

^{1}H NMR spectra of the product were recorded using a 400 MHz FT-NMR spectrometer JNM-A400 (JEOL). The spectrum is shown in Figure 3b. The molar ratio of each monomer unit in the copolymer was 90:1, which was calculated from the peak-integration ratio of peak d and f in the ^{1}H NMR spectrum.

2.3. Preparation of Pure BY Spin-Coated Films

BY was added to DMF (BY concentration: 1.5% by weight). The solution was heated in an oil bath at 150 °C for 4 h and allowed to cool before the undissolved portion was removed by filtration. Pure BY films were prepared by spin-coating the filtered solution onto the UV-O$_3$ treated quartz. The spin-coated films were prepared at 1500 rpm for 30 s. The relative humidity at the time of spin-coating was 14% (RH = 14%), as measured using an RTR-503 temperature and humidity recorder (T&D Corp.). The resulting films were then annealed at 120 °C for 10 min.

2.4. Preparation of BY-PSSV Spin-Coated Films

BY, PSSV, and I2959 were added to mixture of DMF and water. The weight ratio of the components was BY:PSSV:I2959:DMF:water = 1.2×10^1:3.0:6.0×10^{-2}:7.4×10^2:2.2×10^1. The solution was filtered to separate the undissolved portion. The BY-PSSV films were prepared using the filtered solution in the same way as the pure BY film. The resulting films were annealed at 120 °C for 10 min.

2.5. Film Thickness Measurement

Surface roughness was characterized by a white light interferometric BW-S507-N microscope (Nikon Corp., Tokyo, Japan). Bridgelements® was used for the software modules. To measure film thickness, the substrate was exposed by scratching the film with a micro spatula. Subsequently, the height difference between the top and bottom layer of the film was measured by a BW-S507-N microscope. The top-to-bottom height was taken as the film thickness.

2.6. Water Absorption Measurements

The hygroscopicity of pure BY and PSSV was evaluated by the quartz crystal microbalance (QCM) method. The QCM measurements were performed based on the methodology of previous research [50]. As PSSV exhibits high hygroscopicity, the QCM measurements could not guarantee accuracy when RH > 50%. Therefore, the water absorption of PSSV in the high humidity range was evaluated by the following method. Approximately 30 mg of PSSV was exposed to various humidity-controlled environments for several days. Once it reached an equilibrium moisture absorption state, its weight was measured. The relative humidity was controlled using a saturated aqueous solution of inorganic salts such as magnesium nitrate (RH ~50%), sodium chloride (RH ~65%), potassium chloride (RH ~80%), and potassium nitrate (RH ~88%). Experimental values of relative humidity realized in the saturated aqueous solutions of each inorganic salts are given in parentheses. All salts were purchased from Kishida Chemical and used as purchased.

2.7. Photoalignment of the BY Columnar Phase

The spin-coated films were placed on a homemade quartz chamber attached to the RTR-503 humidity sensor. The humidity in the chamber was controlled at approximately 15% using a me-40DP series precise dew-point generator (Micro Equipment). The spin-coated films in the chamber were exposed to linearly polarized visible (LPVis) light (436 nm) passed through a band pass filter and a polarizer using a mercury lamp REX-250 (Asahi Spectra) at room temperature. The light intensity at the sample position was 10 mW cm^{-2}.

Polarized UV-vis absorption spectra were taken on an Agilent 8453 spectrophotometer (Agilent Technologies). The orientation order parameter (S) of BY molecules is defined as $(A_\perp - A_{||})/(A_{\text{large}} + 2A_{\text{small}})$, where $A_\perp$ and $A_{||}$ are absorbances taken with perpendicular and parallel polarized probing beams, respectively. A_{large} and A_{small} represent the larger and smaller absorbances of the two measurements, respectively.

2.8. Evaluation of the Photoaligned BY Columnar Phase by X-ray Scattering Measurements

Grazing-incidence small-angle X-ray scattering (GI-SAXS) measurements were taken by an FR-E X-ray diffractometer equipped with a two-dimensional detector R-axis IV (Rigaku) involving an imaging plate (Fujifilm). An X-ray beam (Cu Kα = 0.154 nm, 0.3 mm collimated) was used, and the camera length was set at 300 mm. The spin-coated films were placed onto a pulse motor stage composed of oblique pulse (ATS-C310-EM, Chuo Precision Industrial) and Z-pulse (ALV-3005-HM, Chuo Precision Industrial) motors. The incident angle of the X-ray beam was adjusted between 0.18 and 0.22° to the substrate surface using the pulse motors.

2.9. UV-Curing of BY Columnar Phase

For UV-curing of the BY columnar phase, non-polarized UV light (365 nm, 5 mW cm^{-2}) passed through band pass filter from a REX-250 was irradiated to the spin-coated films for 5 min.

3. Results and Discussion

3.1. Photoalignment and Humidity Resistance of Pure BY Film

White interference microscopy images of BY films are shown in Figure 1a. The groove in the center of the image was formed when the film was scraped with a micro spatula. A smooth surface morphology with a film thickness of approximately 23 nm was obtained. Figure 1b shows the polarized UV-vis absorption spectrum of the BY film irradiated with LPVis light. The dichroism was induced upon LPVis irradiation. The order parameter (S) for the irradiation dose reached approximately 0.5 at 3 J cm^{-2}, and subsequent irradiation provided S = 0.7 after further dose (Figure 1c), indicating that BY is highly photoaligned by LPVis irradiation.

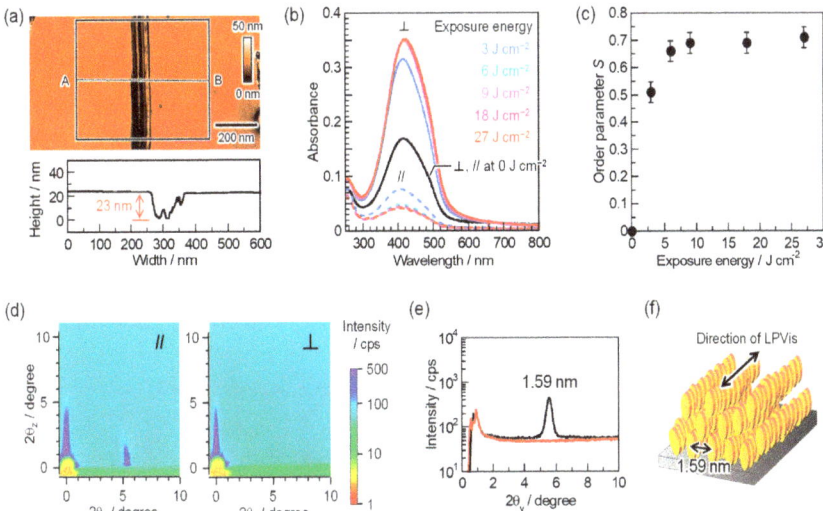

Figure 1. (a) Surface topographical morphology of pure BY film. The cross-section of the average height profile along the A-B line in the box is shown below. (b) Polarized UV-vis absorption spectra change of pure BY films associated with exposure to LPVis light. (c) Order parameter change associated with dose of visible light. (d) GI-SAXS images of pure BY films irradiated with LPVis light of 27 J cm^{-2} dose. (e) In-plane intensity profiles of d. (f) Photoaligned BY columnar phase.

Scattering images and in-plane intensity profiles obtained by GI-SAXS measurements of a BY film irradiated with 27 J cm^{-2} of LPVis light are shown in Figure 1d,e, respectively. When the X-rays were incident parallel to the direction of LPVis light, scattering with an in-plane spacing of 1.59 nm was observed in the in-plane direction, suggesting that the average distance between the columnar aggregates of BY was observed as the scattering peak [39]. Thus, the BY aggregates are uniformly oriented with the column axis parallel to the LPVis light, as shown in Figure 1f.

Figure 2 shows UV-visible spectral changes and S when the photoaligned the BY columnar phase was exposed to a high RH at 90%. A red shift occurred after 1 h of the humidification, and the S value decreased with further humidification. This can be ascribed to the fact that BY exhibits a hygroscopic nature, as shown in Figure 2c. The adsorbed water can cause an orientational relaxation of the hydrated the aggregates, leading to the deterioration of photoalignment.

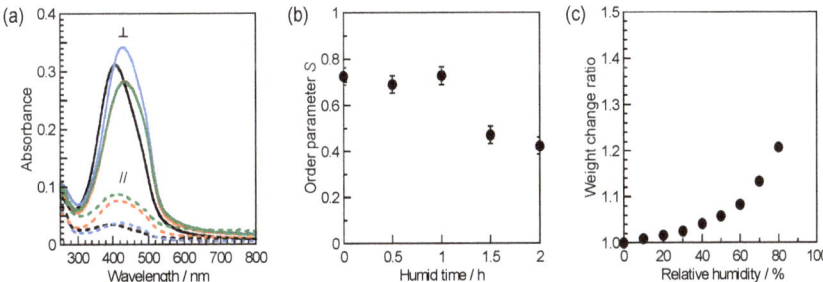

Figure 2. (a) Polarized UV-vis absorption spectra change with humidification at RH = 90% for photoaligned pure BY film. Black, blue, brown, and green lines indicate spectra after humidification of 0, 1, 1.5, and 2 h, respectively. (b) Order parameter change with humidification for photoaligned pure BY film. (c) Humidity-responsive weight change ratios of BY film prepared on a QCM electrode.

3.2. Hygroscopicity of Ionic Linear Polysiloxane Containing Sodium Sulfonate Groups

The ionic linear polysiloxane PSSV was used for the fixation of the BY columnar phase. Spin-coated films of PSSV were prepared on a QCM electrode substrate, and the weight change of the films was monitored upon humidification (Figure 3c). The weight of the films increased as the relative humidity in the chamber increased. Within two minutes of the humidity jump operation, the film reached an equilibrium state. This indicates that the PSSV rapidly absorbs moisture. The humidity dependence of the relative weight on the PSSV in the dry state (RH = 0%) is shown in Figure 3d. Here, the data in the high humidity range at RH > 60% were calculated using the saturated salt method for a bulk PSSV. The weight of PSSV increased continuously with increasing humidity, reaching a factor of approximately 2.2 at RH = 90%. The hygroscopic behavior of PSSV was similar to that of cationic linear polysiloxanes with ammonium salts, which has previously been reported [51].

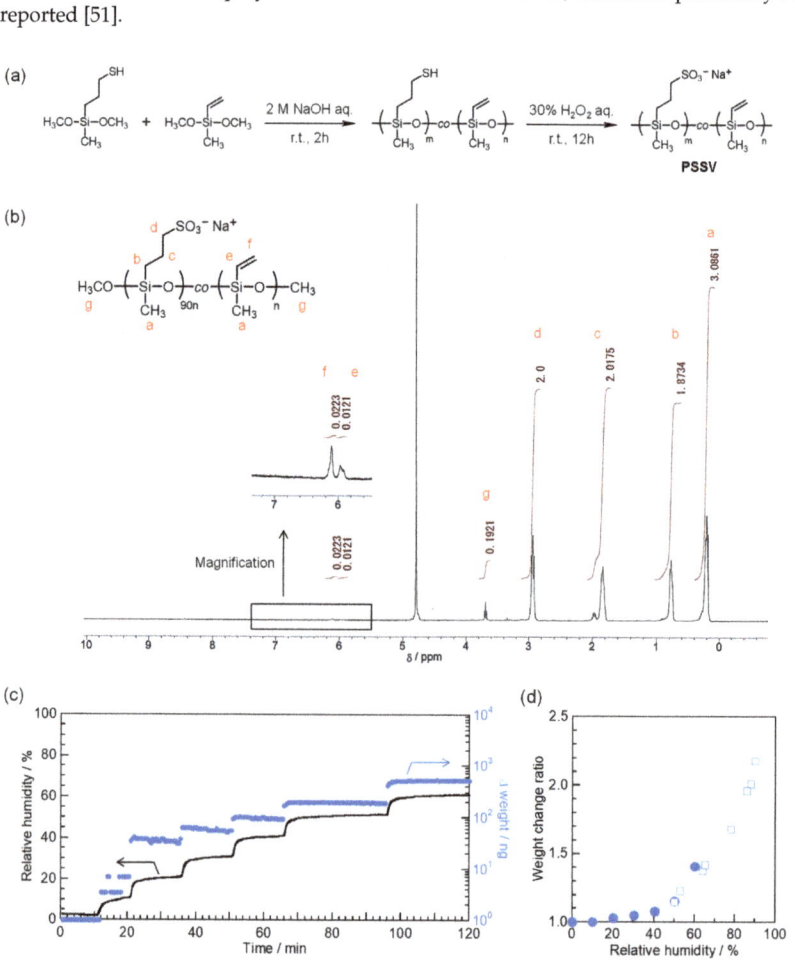

Figure 3. (a) Synthetic scheme to prepare anionic linear polysiloxane PSSV. (b) ^{1}H NMR spectrum of PSSV. (c) Time course profiles of the changes in relative humidity and weight change of PSSV film on QCM electrode. (d) Humidity-responsive weight change ratios of film-state PSSV (circle) and bulk-state PSSV (square).

3.3. Photoalignment of the BY Columnar Phase in PSSV Matrix Using LPVis Light

Figure 4a shows the polarized UV-vis absorption spectra of thin films composed of BY and PSSV when irradiated with LPVis light. The maximum absorption wavelength of the spectrum was blue-shifted by 5 nm from 410 nm by LPVis irradiation, and dichroism was also observed. It is likely that the presence of PSSV prevented BY from forming aggregates before LPVis irradiation, because no blue-shift phenomenon was observed when the pure BY film was irradiated with LPVis light. The S of the films irradiated with polarized visible light above 18 J cm^{-2} was 0.65–0.70 (Figure 4b), indicating that the BY columnar phase was highly photoaligned even in the PSSV matrix. GI-SAXS measurements also yielded data suggesting photoalignment of the BY columnar phase (Figure 4c,d). The periodic structure size after LPVis irradiation was also identical to that of the pure BY film, suggesting that the PSSV does not encompass each column, but bundles the domains of the BY columnar structure as shown in Figure 4e.

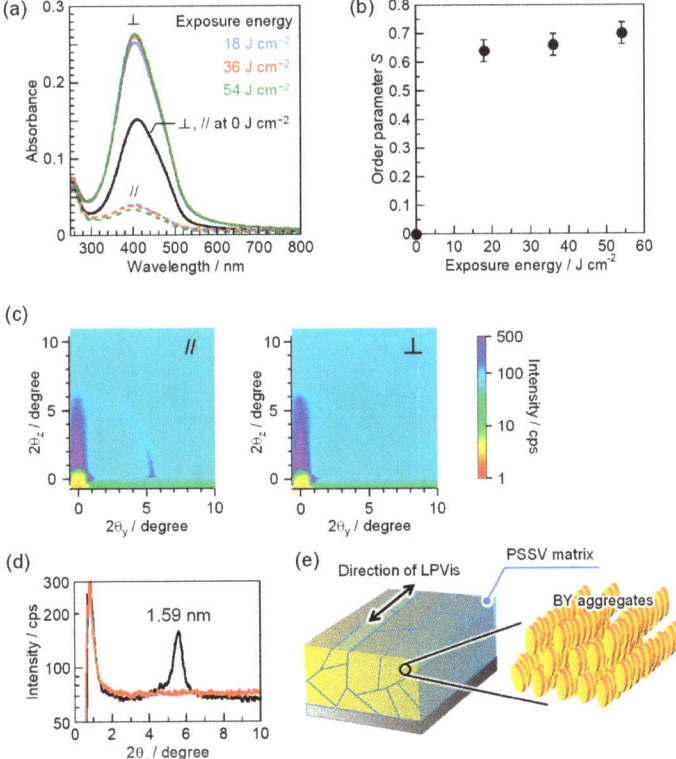

Figure 4. (**a**) Polarized UV-vis absorption spectra change of BY-PSSV films associated with exposure to LPVis light. (**b**) Order parameter change associated with dose of visible light. (**c**) GI-SAXS images of BY-PSSV films irradiated with LPVis light of 36 J cm^{-2} dose. X-ray beams were aligned to two directions parallel and perpendicular to the irradiated light. (**d**) In-plane intensity profiles of c. (**e**) Photoaligned BY columnar phase in PSSV matrix. PSSV is present in the blue region that bundles the BY domain.

3.4. UV Curing of the BY Columnar Phase in the PSSV Matrix

When the photoaligned BY columnar phase within the PSSV matrix was exposed to a RH level of 90% (show in route 1, Figure 5a), a red shift in the ππ* absorption band occurred (Figure 5b). As in the case of the pure BY film, it was considered that the BY columnar phase was hydrated by the absorbed water, leading to the collapse in the photoalignment. The increase in time until S decreases compared to that in the pure BY film is considered to be because the PSSV matrix encapsulates the domains of the BY columnar phase and stabilizes the BY columnar phase. However, the decrease in S was more profound than that of the pure BY film. This may be due to the high hygroscopicity of PSSV, which enables a large amount of water to be incorporated into the film. Once the BY columnar phase begins to collapse, the orientation order is suddenly reduced after two hours. The GI-SAXS profiles of the film exposed to humidification also showed a peak (d = 1.91 nm) when the X-ray was incident perpendicular to the LPVis, suggesting a randomization of the photoaligned BY columnar phase. The peak position differs from that before humidification (Figure 4d) because the BY aggregates' transition takes place from the nematic columnar phase to the rectangular columnar phase due to humidification [39].

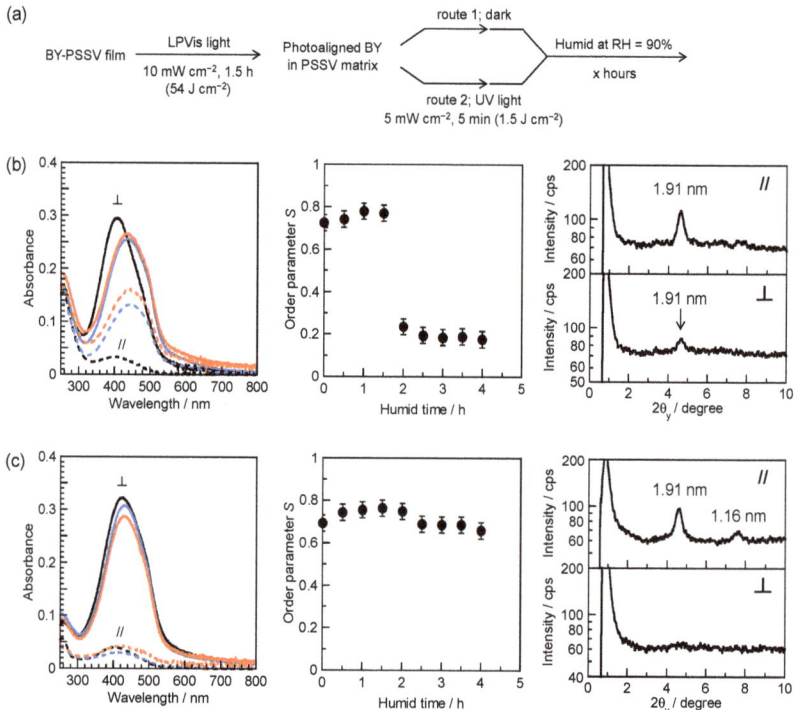

Figure 5. (**a**) Schematic procedure for BY-PSSV film. (**b**) Polarized UV-vis absorption spectra change (**left**), order parameter change (**center**), and in-plane profiles obtained by GI-SAXS measurements (**right**) with humidification at RH = 90% for BY-PSSV film passing through route 1. In the left figure of (**b**), black, blue, and brown lines indicate spectra after humidification of 0, 2, and 4 h, respectively. The right side profiles in (**b**) were obtained by GI-SAXS measurements after humidification of 2 h. (**c**) Data of BY-PSSV film passing through route 2, corresponding to (**b**).

Conversely, S was retained under the same humidity conditions when the UV-irradiated BY-PSSV film was exposed to a humid environment (route 2 in Figure 5a). Additionally, GI-SAXS measurements indicated that the photoalignment of the BY columnar phase was retained (Figure 5c). Thus, it was found that the photoaligned BY columnar phase could be firmly fixed by the crosslinking of PSSV. Figure 6a shows a visual photograph of the BY-PSSV film prepared on a quartz substrate after irradiation with LPVis and non-polarized UV lights, followed by further humidification at RH = 90% for 4 h. Figure 6b displays a photograph of the same film in Figure 6a, taken through a polarizer set above the film. The color contrast of the film was significantly different when the direction of the transmitted light was perpendicular or parallel to direction of the LPVis light. In Figure 6c, the surface morphology of the film is shown. Some dotted protrusions were observed when compared to one of the pure BY film (see Figure 1a); however, overall, the surface of the BY-PSSV film was flat. These results indicate that it was possible to form the films on centimeter-scale substrates and to prepare large-area dichroic films even after humidification. Combined with pattern exposure through a photomask, various absorption anisotropic films can be developed. The film defects observed in Figure 6b were caused by repeatedly grabbing the film using tweezers during each humidification procedure and spectroscopic measurements.

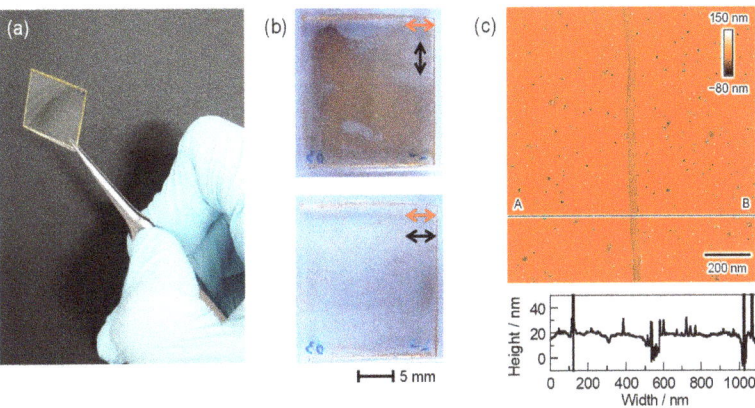

Figure 6. (**a**) Snapshot of BY-PSSV film on quartz. (**b**) Observation of BY-PSSV film through polarizer. The film was irradiated with an LPVis light of 36 J cm^{-2} dose and a UV light of 1.5 J cm^{-2} dose, then exposed to a humid atmosphere of RH = 90% for 4 h. Brown and black arrows indicate the directions of exposed LPVis light and polarizer, respectively. (**c**) Surface topographical morphology of the BY-PSSV film by white light interference microscopy. The cross-section of the average height profile along the A-B line in the top view image is shown below.

4. Conclusions

In this study, photoalignment and photo-fixation of the BY columnar phase were achieved using anionic linear polysiloxane PSSV as a matrix. Using light of two different wavelengths, the photoalignment and photo-fixation were independently achieved. PSSV bundles the domain of the BY columnar phase; thus, after the UV curing, the film dichroism remained as high as that of pure BY films, even after long-term exposure to the high-humidity environment. As the BY columnar phase is often used as photoalignment sublayers for low-molecular-mass nematic liquid crystals, this method, which improves the moisture resistance, is expected to expand the applications of BY. The BY alignment layer is also expected to be used as the alignment sublayer for water-soluble compounds. This proposal is expected to expand the possibilities of anisotropic materials for the alignment of water-soluble polymers, and for inducing molecular orientation of molecular aggregates in water media.

Author Contributions: M.H. and T.S. conceived and designed the project. A.M. conducted most of the experiments. S.N. provided experimental technical assistance and discussions on the data. M.H. and T.S. wrote the manuscript. All authors have read and agreed to the published version of the manuscript.

Funding: This work was supported by JSPS KAKENHI Grant Numbers JP18K14283, JP20H05217, JP22H02142, and JP22H04536 for MH, and JP21H01983 and JP21K19000 for TS. MH is also thankful for financial support from the Toshiaki Ogasawara Memorial Foundation, the Asahi Glass Foundation, and the Toukai Foundation for Technology.

Data Availability Statement: Not applicable.

Acknowledgments: We would like to thank Tatsuo Hikage of Nagoya University for their assistance with GI-SAXS measurements.

Conflicts of Interest: The authors declare no conflict of interest.

References

1. Kato, T.; Mizoshita, N.; Kishimoto, K. Functional Liquid-Crystalline Assemblies: Self-Organized Soft Materials. *Angew. Chem. Int. Ed.* **2006**, *45*, 38–68. [CrossRef] [PubMed]
2. Bukusoglu, E.; Bedolla Pantoja, M.; Mushenheim, P.C.; Wang, X.; Abbott, N.L. Design of Responsive and Active (Soft) Materials Using Liquid Crystals. *Annu. Rev. Chem. Biomol. Eng.* **2016**, *7*, 163–196. [CrossRef] [PubMed]
3. Kloos, J.; Joosten, N.; Schenning, A.; Nijmeijer, K. Self-assembling liquid crystals as building blocks to design nanoporous membranes suitable for molecular separations. *J. Membr. Sci.* **2021**, *620*, 118849. [CrossRef]
4. Uchida, J.; Soberats, B.; Gupta, M.; Kato, T. Advanced Functional Liquid Crystals. *Adv. Mater.* **2022**, *34*, 2109063. [CrossRef]
5. Bisoyi, H.K.; Li, Q. Liquid Crystals: Versatile Self-Organized Smart Soft Materials. *Chem. Rev.* **2022**, *122*, 4887–4926. [CrossRef]
6. Tam-Chang, S.-W.; Huang, L. Chromonic liquid crystals: Properties and applications as functional materials. *Chem. Commun.* **2008**, *17*, 1957–1967. [CrossRef]
7. Lydon, J. Chromonic review. *J. Mater. Chem.* **2010**, *20*, 10071–10099. [CrossRef]
8. Collings, P.J.; Goldstein, J.N.; Hamilton, E.J.; Mercado, B.R.; Nieser, K.J.; Regan, M.H. The nature of the assembly process in chromonic liquid crystals. *Liq. Cryst. Rev.* **2015**, *3*, 1–27. [CrossRef]
9. Dierking, I.; Martins Figueiredo Neto, A. Novel Trends in Lyotropic Liquid Crystals. *Crystals* **2020**, *10*, 604. [CrossRef]
10. Bosire, R.; Ndaya, D.; Kasi, R.M. Recent progress in functional materials from lyotropic chromonic liquid crystals. *Polym. Int.* **2021**, *70*, 938–943. [CrossRef]
11. Ichikawa, T.; Kuwana, M.; Suda, K. Chromonic Ionic Liquid Crystals Forming Nematic and Hexagonal Columnar Phases. *Crystals* **2022**, *12*, 1548. [CrossRef]
12. Schneider, T.; Lavrentovich, O.D. Self-Assembled Monolayers and Multilayered Stacks of Lyotropic Chromonic Liquid Crystalline Dyes with In-Plane Orientational Order. *Langmuir* **2000**, *16*, 5227–5230. [CrossRef]
13. Iverson, I.K.; Casey, S.M.; Seo, W.; Tam-Chang, S.-W.; Pindzola, B.A. Controlling Molecular Orientation in Solid Films via Self-Organization in the Liquid-Crystalline Phase. *Langmuir* **2002**, *18*, 3510–3516. [CrossRef]
14. Hara, M.; Nagano, S.; Mizoshita, N.; Seki, T. Chromonic/Silica Nanohybrids: Synthesis and Macroscopic Alignment. *Langmuir* **2007**, *23*, 12350–12355. [CrossRef] [PubMed]
15. Park, S.-K.; Kim, S.-E.; Kim, D.-Y.; Kang, S.-W.; Shin, S.; Kuo, S.-W.; Hwang, S.-H.; Lee, S.H.; Lee, M.-H.; Jeong, K.-U. Polymer-Stabilized Chromonic Liquid-Crystalline Polarizer. *Adv. Funct. Mater.* **2011**, *21*, 2129–2139. [CrossRef]
16. Ichimura, K. Photoalignment of Liquid-Crystal Systems. *Chem. Rev.* **2000**, *100*, 1847–1873. [CrossRef]
17. Seki, T.; Nagano, S.; Hara, M. Versatility of photoalignment techniques: From nematics to a wide range of functional materials. *Polymer* **2013**, *54*, 6053–6072. [CrossRef]
18. Priimagi, A.; Barrett, C.J.; Shishido, A. Recent twists in photoactuation and photoalignment control. *J. Mater. Chem. C* **2014**, *2*, 7155–7162. [CrossRef]
19. Ichimura, K.; Fujiwara, T.; Momose, M.; Matsunaga, D. Surface-assisted photoalignment control of lyotropic liquid crystals. Part 1. Characterization and photoalignment of aqueous solutions of a water-soluble dye as lyotropic liquid crystals. *J. Mater. Chem.* **2002**, *12*, 3380–3386. [CrossRef]
20. Fujiwara, T.; Ichimura, K. Surface-assisted photoalignment control of lyotropic liquid crystals. Part 2. Photopatterning of aqueous solutions of a water-soluble anti-asthmatic drug as lyotropic liquid crystals. *J. Mater. Chem.* **2002**, *12*, 3387–3391. [CrossRef]
21. Hara, M.; Nagano, S.; Kawatsuki, N.; Seki, T. Photoalignment and patterning of a chromonic-silica nanohybrid on photocrosslinkable polymer thin films. *J. Mater. Chem.* **2008**, *18*, 3259–3263. [CrossRef]
22. Van der Asdonk, P.; Hendrikse, H.C.; Fernandez-Castano Romera, M.; Voerman, D.; Ramakers, B.E.I.; Löwik, D.W.P.M.; Sijbesma, R.P.; Kouwer, P.H.J. Patterning of Soft Matter across Multiple Length Scales. *Adv. Funct. Mater.* **2016**, *26*, 2609–2616. [CrossRef]
23. Guo, Y.; Jiang, M.; Peng, C.; Sun, K.; Yaroshchuk, O.; Lavrentovich, O.D.; Wei, Q.-H. Designs of Plasmonic Metamasks for Photopatterning Molecular Orientations in Liquid Crystals. *Crystals* **2017**, *7*, 8. [CrossRef]
24. Wani, O.M.; Zeng, H.; Wasylczyk, P.; Priimagi, A. Programming Photoresponse in Liquid Crystal Polymer Actuators with Laser Projector. *Adv. Opt. Mater.* **2018**, *6*, 1700949. [CrossRef]
25. Wang, J.; McGinty, C.; Reich, R.; Finnemeyer, V.; Clark, H.; Berry, S.; Bos, P. Process for a Reactive Monomer Alignment Layer for Liquid Crystals Formed on an Azodye Sublayer. *Materials* **2018**, *11*, 1195. [CrossRef]
26. Berteloot, B.; Nys, I.; Poy, G.; Beeckman, J.; Neyts, K. Ring-shaped liquid crystal structures through patterned planar photoalignment. *Soft Matter* **2020**, *16*, 4999–5008. [CrossRef] [PubMed]
27. Yin, K.; Xiong, J.; He, Z.; Wu, S.-T. Patterning Liquid-Crystal Alignment for Ultrathin Flat Optics. *ACS Omega* **2020**, *5*, 31485–31489. [CrossRef]
28. Kim, K.; Kim, S.-U.; Choi, S.; Heo, K.; Ahn, S.-K.; Na, J.-H. High-Definition Optophysical Image Construction Using Mosaics of Pixelated Wrinkles. *Adv. Sci.* **2020**, *7*, 2002134. [CrossRef]

29. Nys, I.; Berteloot, B.; Poy, G. Surface Stabilized Topological Solitons in Nematic Liquid Crystals. *Crystals* **2020**, *10*, 840. [CrossRef]
30. Xiong, J.; Wu, S.-T. Planar liquid crystal polarization optics for augmented reality and virtual reality: From fundamentals to applications. *eLight* **2021**, *1*, 3. [CrossRef]
31. Kim, H.; Abdelrahman, M.K.; Choi, J.; Kim, H.; Maeng, J.; Wang, S.; Javed, M.; Rivera-Tarazona, L.K.; Lee, H.; Ko, S.H.; et al. From Chaos to Control: Programmable Crack Patterning with Molecular Order in Polymer Substrates. *Adv. Mater.* **2021**, *33*, 2008434. [CrossRef] [PubMed]
32. Padmini, H.N.; Rajabi, M.; Shiyanovskii, S.V.; Lavrentovich, O.D. Azimuthal Anchoring Strength in Photopatterned Alignment of a Nematic. *Crystals* **2021**, *11*, 675. [CrossRef]
33. Folwill, Y.; Zeitouny, Z.; Lall, J.; Zappe, H. A practical guide to versatile photoalignment of azobenzenes. *Liq. Cryst.* **2021**, *48*, 862–872. [CrossRef]
34. McGinty, C.P.; Kołacz, J.; Spillmann, C.M. Large rewritable liquid crystal pretilt angle by in situ photoalignment of brilliant yellow films. *Appl. Phys. Lett.* **2021**, *119*, 141111. [CrossRef]
35. Liu, S.; Nys, I.; Neyts, K. Two-Step Photoalignment with High Resolution for the Alignment of Blue Phase Liquid Crystal. *Adv. Opt. Mater.* **2022**, *10*, 2200711. [CrossRef]
36. Pinchin, N.P.; Lin, C.-H.; Kinane, C.A.; Yamada, N.; Pena-Francesch, A.; Shahsavan, H. Plasticized liquid crystal networks and chemical motors for the active control of power transmission in mechanical devices. *Soft Matter* **2022**, *18*, 8063–8070. [CrossRef] [PubMed]
37. Lall, J.; Zappe, H. In situ, spatially variable photoalignment of liquid crystals inside a glass cell using brilliant yellow. In Proceedings of the Photosensitive Materials and their Applications II, Strasbourg, France, 3 April–23 May 2022; Volume 12151.
38. Li, Y.; Luo, Z.; Wu, S.-T. High-Precision Beam Angle Expander Based on Polymeric Liquid Crystal Polarization Lenses for LiDAR Applications. *Crystals* **2022**, *12*, 349. [CrossRef]
39. Matsumori, M.; Takahashi, A.; Tomioka, Y.; Hikima, T.; Takata, M.; Kajitani, T.; Fukushima, T. Photoalignment of an Azobenzene-Based Chromonic Liquid Crystal Dispersed in Triacetyl Cellulose: Single-Layer Alignment Films with an Exceptionally High Order Parameter. *ACS Appl. Mater. Interfaces* **2015**, *7*, 11074–11078. [CrossRef] [PubMed]
40. Wang, J.; McGinty, C.; West, J.; Bryant, D.; Finnemeyer, V.; Reich, R.; Berry, S.; Clark, H.; Yaroshchuk, O.; Bos, P. Effect of humidity and surface on photoalignment of brilliant yellow. *Liq. Cryst.* **2016**, *44*, 863–872. [CrossRef]
41. Shi, Y.; Zhao, C.; Ho, J.Y.-L.; Vashchenko, V.V.; Srivastava, A.K.; Chigrinov, V.G.; Kwok, H.-S.; Song, F.; Luo, D. Exotic Property of Azobenzenesulfonic Photoalignment Material Based on Relative Humidity. *Langmuir* **2017**, *33*, 3968–3974. [CrossRef]
42. Shi, Y.; Zhao, C.; Ho, J.Y.-L.; Song, F.; Chigrinov, V.G.; Luo, D.; Kwok, H.-S.; Sun, X.W. High Photoinduced Ordering and Controllable Photostability of Hydrophilic Azobenzene Material Based on Relative Humidity. *Langmuir* **2018**, *34*, 4465–4472. [CrossRef] [PubMed]
43. Wan, Y.; Zhao, D. On the Controllable Soft-Templating Approach to Mesoporous Silicates. *Chem. Rev.* **2007**, *107*, 2821–2860. [CrossRef] [PubMed]
44. Hara, M. Mesostructure and orientation control of lyotropic liquid crystals in a polysiloxane matrix. *Polym. J.* **2019**, *51*, 989–996. [CrossRef]
45. Huo, Q.; Margolese, D.I.; Ciesla, U.; Feng, P.; Gier, T.E.; Sieger, P.; Leon, R.; Petroff, P.M.; Schüth, F.; Stucky, G.D. Generalized synthesis of periodic surfactant/inorganic composite materials. *Nature* **1994**, *368*, 317–321. [CrossRef]
46. Hara, M.; Orito, T.; Nagano, S.; Seki, T. Humidity-responsive phase transition and on-demand UV-curing in a hygroscopic polysiloxane-surfactant nanohybrid film. *Chem. Commun.* **2018**, *54*, 1457–1460. [CrossRef]
47. Hara, M.; Wakitani, N.; Kodama, A.; Nagano, S.; Seki, T. Hierarchical Photocomposition of Heteronanostructures in a Surfactant-Polysiloxane Hybrid Film toward Next-Generation Nanolithography. *ACS Appl. Polym. Mater.* **2020**, *2*, 2284–2290. [CrossRef]
48. Kaneko, Y. Ionic silsesquioxanes: Preparation, structure control, characterization, and applixations. *Polymer* **2018**, *144*, 205–224. [CrossRef]
49. Kaneko, Y.; Toyodome, H.; Mizumo, T.; Shikinaka, K.; Iyi, N. Preparation of a Sulfo-Group-Containing Rod-Like Polysilsesquioxane with a Hexagonally Stacked Structure and Its Proton Conductivity. *Chem. Eur. J.* **2014**, *20*, 9394–9399. [CrossRef]
50. Hara, M.; Ueno, Y.; Nagano, S.; Seki, T. Water-Retentive/Lipophilic Amphiphilic Surface Properties Attained by Hygroscopic Polysiloxane Ultrathin Films. *J. Fiber Sci. Technol.* **2022**, *78*, 169–177. [CrossRef]
51. Hara, M.; Iijima, Y.; Nagano, S.; Seki, T. Simple linear ionic polysiloxane showing unexpected nanostructure and mechanical properties. *Sci. Rep.* **2021**, *11*, 17683. [CrossRef]

Disclaimer/Publisher's Note: The statements, opinions and data contained in all publications are solely those of the individual author(s) and contributor(s) and not of MDPI and/or the editor(s). MDPI and/or the editor(s) disclaim responsibility for any injury to people or property resulting from any ideas, methods, instructions or products referred to in the content.

Communication

Angular Dependence of Guest–Host Liquid Crystal Devices with High Pretilt Angle Using Mixture of Vertical and Horizontal Alignment Materials

Masahiro Ito [1,*], Eriko Fukuda [2], Mitsuhiro Akimoto [3], Hikaru Hoketsu [3], Yukitaka Nakazono [3], Haruki Tohriyama [3] and Kohki Takatoh [3]

1. Department of Medical Course, Faculty of Health and Medical Science, Teikyo Heisei University, Tokyo 170-8445, Japan
2. Department of Electrical Engineering, Faculty of Science and Engineering, Kyushu Sangyo University, Fukuoka 813-8503, Japan; e.fukuda@ip.kyusan-u.ac.jp
3. Department of Electrical Engineering, Faculty of Engineering, Sanyo-Onoda City University, Yamaguchi 756-0884, Japan; mt-akimoto@rs.socu.ac.jp (M.A.); takaoh@rs.socu.ac.jp (K.T.)
* Correspondence: masahiro.ito@thu.ac.jp; Tel.: +81-3-5843-3072

Abstract: To date, devices exhibiting incidence-angle-dependent transmittance have been fabricated by imparting an angle to a bulk liquid crystal (LC) by aligning the LC in the vicinity of one substrate horizontally (with respect to the substrate) while aligning the LC in the vicinity of another substrate vertically. Another approach has been to control LC angles near substrates by blending or layering horizontal and vertical alignment films. In this study, we control LC angles near substrates by controlling the pretilt angles of blended alignment films; for specific angles, we use dichroic dyes to characterize the incidence angle dependence of these LC devices. Using a guest/host LC device with a pretilt angle near 45°, we successfully construct an LC element with a transmittance peak near a polar angle of 45°.

Keywords: liquid crystal; rubbing method; composite alignment layer; high pretilt angle; incidence angle dependence

1. Introduction

Louver is a popular tool to control the incoming light by opening and closing its panels mechanically. By opening the panels halfway, light can enter on one side and be blocked on the other side. The adjustability of the angle of incoming light is the most distinctive feature of louvers. That is, it can block light coming in from above, such as sunlight, while allowing the view parallel to or below the line of sight to be observed. In view of the recent demands on the smartification of building materials, motionless and electrically active optical filters that reduce the transmittance of light in one direction and increase transmittance in the front or the other direction are highly desirable.

Liquid crystals doped with dichroic dyes are one of the candidates for such a 'smart' louver. Dichroic dye molecules are aligned parallel to the molecular axis of the liquid crystal and absorb polarized light that vibrates parallel to the molecular axis. Devices with a unique angular dependence were demonstrated by control of liquid crystal (LC) polar angles. Among them, a device that exhibits a difference in transmittance depending on whether the angle of incidence is positive or negative was developed using a hybrid alignment nematic (HAN) LC [1], a polymer dispersed LC (PDLC) [2,3], a hybrid twisted nematic (HTN) LC [4], and a guest–host HAN LC [5]. The louver function can be achieved using HAN-type liquid crystals, but it is necessary to apply a voltage to control the angle with the highest transmittance. If the angle of incidence is specified, it is convenient to angle the liquid crystal with no voltage applied. A method to control the pretilt angle of the liquid crystal makes this possible, which is the purpose of this study. This is made possible by controlling the initial angle of the liquid crystal by preparing the blend ratio of horizontally and vertically aligned films. It is also possible to maximize transmittance in the direction perpendicular to the substrate, as in normal liquid crystal devices, by applying a voltage.

The same properties have also been obtained by electrically controlled birefringence (ECB) devices with a high pretilt angle (the angle of LCs in the vicinity of the alignment layer) [5]. The alignment is controlled by strain at the alignment layer surface, which in turn depends on the stress applied to the alignment film during the rubbing process. The direction of the strain is also closely related to the pretilt angle [6]. The pretilt angles were also controlled by using a method in which the monomers are polymerized by UV irradiation under an applied voltage (such as polymer sustained alignment [7–9] or nano-phase-separated [10,11]), regulating the baking temperature [12–14], adjusting the rubbing strength [14,15], using a mixture of vertical and horizontal alignment materials [16,17], or stacking a vertical (horizontal) alignment material on a horizontal (vertical) alignment material [15,18,19].

Horizontal and vertical alignment films are typically not miscible in solid configurations, so a segregation process takes place as liquid films are dried. The precipitation speed, the relative solubility in a solvent, and the surface properties of the polyimides all play key roles in determining the final structure of the film. Yeung et al. used JALS9203 and JALS2021 (Japan Synthetic Rubber Co., Ltd., Tokyo, Japan) to apply coatings to glass substrates and reported that nanostructured domains formed naturally upon drying [20]. The final pretilt angle depends on the area ratio of the horizontal and vertical domains, the relative anchor strength of the domains, and the elastic constants for the LCs.

In the present study, we use blended alignment films formed by varying the blending ratio of vertical and horizontal alignment films. We also control the pretilt angle by varying the rubbing strength and baking temperature, characterizing pretilt angles at smaller sizes than those considered by Yeung et al. We measure pretilt angles at 160 μm intervals over an area of 10.2 cm^2 using an OPTIPRO-micro (Shintech Inc., Yamaguchi, Japan) with a laser spot size of 3 μm.

The rubbing strength (RS) is given by:

$$RS = Nl\left(1 + \frac{2\pi rn}{60v}\right) \quad (1)$$

where N is the number of rubbings; l is the pile contact length (mm); r is roller radius, pile thickness (mm); n is the roller revolution speed (rpm); and v is the movement speed of the stage (mm/s) [21].

The pretilt angle for an LC device was measured by a pretilt analysis system using the crystal rotation method [22].

$$I = \frac{1}{2}\sin^2\frac{\Gamma}{2} \quad (2)$$

$$\Gamma = \frac{2\pi}{\lambda}d\left\{\frac{1}{c^2}\left(a^2 - b^2\right)\sin\theta_p\cos\theta_p\sin\phi + \frac{1}{c}\sqrt{1 - \frac{a^2b^2}{c^2}\sin^2\phi} - \frac{1}{b}\sqrt{1 - b^2\sin^2\phi}\right\} \quad (3)$$

$$a = \frac{1}{n_e}, b = \frac{1}{n_o}, c^2 = a^2\cos\theta_p + b^2\sin\theta_p \quad (4)$$

where I is the amount of incident light, n_e is the extraordinary index, n_o is the ordinary index, ϕ is the angle of the cell, θ_p is the pretilt angle, and d is the cell thickness.

For the pretilt angle mapping measurement, the optical axis of one polarizer was set at 45° with respect to one of the rubbing directions in a crossed-nicols configuration, and the wavelength was 550 nm. The pretilt angle mappings were measured using an OPTIPRO-micro. The Stokes parameter S_2 is defined as the difference between the optical intensity transmitted by a linear polarizer oriented 45° to the x-axis ($I_{45°}$) and one oriented 135° ($I_{135°}$). The Stokes parameter S_3 is the difference between the left and right circularly polarized power. These parameters can also be described by the retardation δ and E_x, as well as E_y, which are the flux density transmitted by a linear polarizer oriented parallel to the x-axis and that transmitted by one oriented parallel to the y-axis, respectively.

$$S_2(\theta) = I_{45°} - I_{135°} = 2|E_x||E_y|\cos\delta \tag{5}$$

$$S_3(\theta) = I_L - I_R = 2|E_x||E_y|\sin\delta \tag{6}$$

where θ is the incidence angle. The retardation δ is calculated by:

$$\delta = \Delta n' d = \{n'_e(\theta) - n'_o(\theta)\}d \tag{7}$$

$$n'_e(\theta) = -\frac{\varepsilon_{xz}}{\varepsilon_{zz}} n \sin\theta + \sqrt{\frac{n_e^2 n_o^2}{\varepsilon_{zz}} - \frac{\varepsilon_{xx}\varepsilon_{zz} - \varepsilon_{xz}}{\varepsilon_{zz}^2}(n\sin\theta)^2} \tag{8}$$

$$n'_o(\theta) = \sqrt{n_o^2 - (n\sin\theta)^2} \tag{9}$$

$$\varepsilon_{zz} = n_o^2 + \left(n_e^2 - n_o^2\right)\sin^2\theta_{LC} \tag{10}$$

$$\varepsilon_{xx} = n_o^2 + \left(n_e^2 - n_o^2\right)\cos^2\theta_{LC}\cos^2\alpha^{in} \tag{11}$$

$$\varepsilon_{xz} = \left(n_e^2 - n_o^2\right)\sin\theta_{LC}\cos\theta_{LC}\cos^2\alpha^{in} \tag{12}$$

where d is the thickness; n_e, and n_o are refractive indices; n is the refractive index of air; ε_{xx}, ε_{xz}, and ε_{zz} are permittivity tensors; and αin is the azimuth angle. The average tilt angle θ_{LC} was estimated from the retardation δ.

After characterizing LC devices with large pretilt angles using blended alignment films, we measured the incidence angle dependence by adding dichroic dyes to LCs and injecting these mixtures into cells with two different pretilt angles.

2. Materials and Methods

2.1. Pretilt Angle Preparations

The alignment films adopted were horizontal polyimide (H-PI) PIA-X359-01X (Chisso Petrochemical Corp., Tokyo, Japan) and SE-150 (Nissan Chemical, Tokyo, Japan) and vertical polyimide (V-PI) PIA-X768-01X (Chisso Petrochemical Corp., Tokyo Japan) and SE-4811 (Nissan Chemical, Tokyo, Japan). To obtain various pretilt angles, the concentration ratios of H-PI to V-PI, the baking temperature, and the rubbing strength were controlled. The diluents were the same for PIA-X359-01X and PIA-X768-01X and for SE-150 and SE-4811. The composite PI was coated on a substrate by spin coating and then baked and rubbed. Two substrates that had been treated identically were assembled to produce an empty anti-parallel LC cell with a 20 µm gap using silica bead spacers. The LC materials ZLI-4792 (Merck & Co., Inc., Darmstadt, Germany) [23] and MLC-2038 (Merck & Co., Inc., Darmstadt, Germany) were used. ZLI-4792 and MLC-2038 have positive dielectric anisotropy (5.3) and negative dielectric anisotropy (−5.0), respectively. The pretilt angles for an LC device were measured by a pretilt analysis system (PAS-301, Elsicon Inc., Newark, NJ, USA) using the crystal rotation method, and pretilt angle mappings were measured by an OPTIPRO micro (Shintech Inc., Yamaguchi, Japan). We can observe averaged optical pretilt angles within a range of 3 µm in diameter using the OPTIPRO micro.

2.2. High Pretilt Angle LCD Preparation

The cells were prepared with various polyimide composites of H-PI and V-PI, which gave three pretilt angles (22, 44, and 48°). Two substrates that had been treated identically were assembled to produce an empty anti-parallel LC cell with a 5.0 μm gap using silica bead spacers. The mixtures (guest–host LC: GHLC) of LC material ZLI-4792 and a dichroic dye NKX-4173 (Hayashibara Ltd., Okayama, Japan) were injected at concentrations of 5 wt%. For obtaining the dependence between the direction of the incidence angles and the polar angle of the LC molecules, the incidence angle dependence of the transmittance of the LCDs with polarizers was measured using an optical property measurement system (RETS-1100, Otsuka Electronics Corp., Osaka, Japan). All measurements were carried out using light with a wavelength of 550 nm.

Simulations of the optical properties were performed using LCD Master (Shintech Inc., Yamaguchi, Japan) software under two kinds of pretilt angles under a cell thickness of 5 μm for GHLC (the concentration of dichroic dye was 5 wt%). The optical properties were calculated using the extended Jones matrix method with 2×2.

3. Results and Discussion

3.1. Pretilt Angle

3.1.1. Pretilt Angles for Individual PI Materials

The solid points in Figure 1 show pretilt angles for ZLI-4792 using various RS values, while the open squares show those for MLC-2038. The squares, circle, inverse triangle, and triangle indicate pretilt angles for RS values of 73.0, 49.8, 35.3, and 0 mm, respectively. The pretilt angle differs depending on the PI. This is expected to be due to differences in surface tension [24]. When vertical alignment films and LCs with positive dielectric anisotropy were injected, the pretilt angle increased with decreasing RS. With no rubbing, the pretilt angles remained in the vicinity of 90°. When using SE-4811, the use of LCs with negative dielectric anisotropy yielded pretilt angles near 90° even for an RS of 73 mm. In contrast, when using PIA-X768-01X, we found a pretilt angle near 85° for an RS of 35.3 mm, but this decreased to around 55° when RS was increased. This decrease in pretilt angle is thought to be because PIA-X768-01X is sensitive to rubbing.

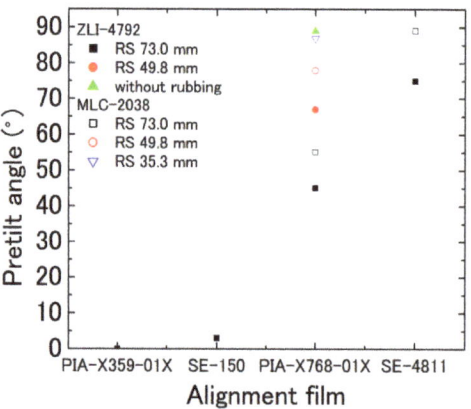

Figure 1. Pretilt angles for ZLI-4792 or MLC-2038 for different PI films as function of RS.

3.1.2. Pretilt Angles for Composite PIA-X359-01X and PIA-X768-01X

The polyimides were baked at 220 °C because this is the baking temperature for PIA-X359-01X and PIA-X768-01X. Figure 2a shows the pretilt angle as a function of wt% PIA-X768-01X in PIA-X359-01X for different RS values. It can be seen that any pretilt angle can be generated using this method. The pretilt angles of PIA-X359-01X and PIA-X768-01X (blue inverse triangles) are 0 and 90° when RS is zero, respectively. Black squares, red circles, and green triangles show RS of 73.0 mm, 49.8 mm, and 35.3 mm, respectively. The pretilt angle has a linear relationship up to 70 wt% and then saturates. For lower RS, the linear relationship is maintained up to high concentrations. Figure 3a shows the appearance of a typical cell under the crossed-nicols configuration (the optical axis of one polarizer set at 45° with respect to one of the rubbing directions). The vertical axis in Figure 3 is the rubbing strength, and the horizontal axis is wt% PIA-X768-01X in PIA-X359-01X. The rubbing stripes were generated for pretilt angles of more than 20°. We suspect that when the concentration of the vertical alignment film is high and the RS is large, the film cannot withstand the pressure and the horizontal alignment film becomes dominant, causing the pretilt angle to saturate. Further evidence for this interpretation comes from the fact that, in Figure 1, the pretilt angle decreases upon rubbing of PIA-X768-01X.

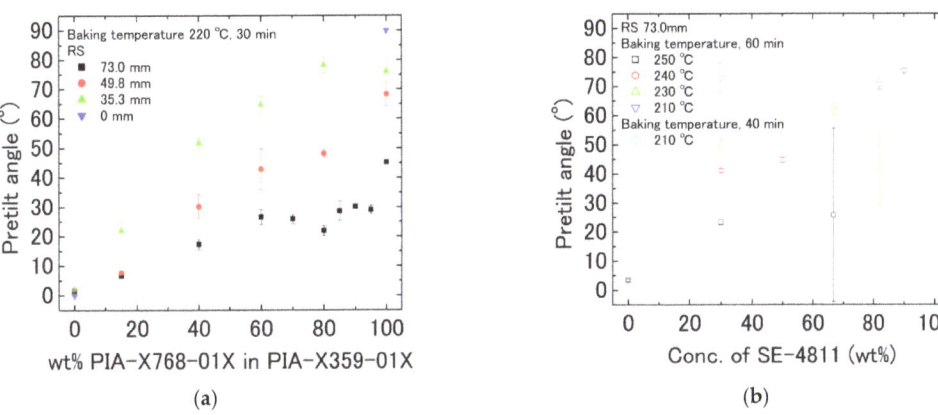

Figure 2. Pretilt angle vs. relative concentration of vertical to horizontal alignment film for (**a**) PIA-X768-01X and PIA-X359-01X and (**b**) SE-4811 and SE-150.

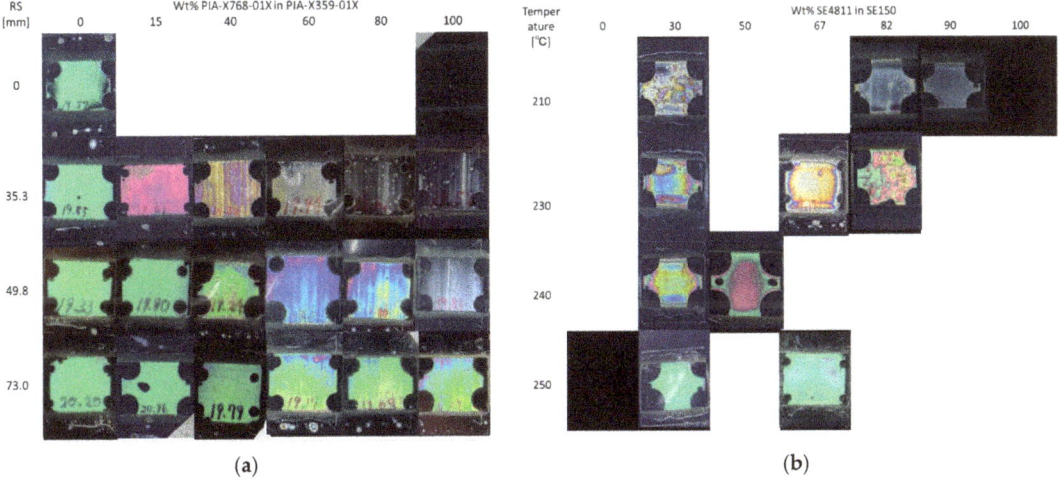

Figure 3. Cell photographs of blended alignment films with polarizers mounted on crossed nicols (45–135° relative to rubbing direction) for (**a**) PIA-X768-01X and PIA-X359-01X and (**b**) SE-4811 and SE-150.

3.1.3. Pretilt Angles for Composite SE-150 and SE-4811

The baking temperatures for SE-150 and SE-4811 are 250 °C and 210 °C, respectively. All rubbing strengths were 73.0 mm. Figure 2b shows the pretilt angle as a function of wt% SE-4811 in SE-150 for different RS values. The pretilt angles of SE150 (black open squares) and SE-4811 (sky-blue open diamonds) are 4 and 75° when RS is zero, respectively. Black open squares, red open circles, green open triangles, and blue open inversed triangles show temperatures of 250 °C, 240 °C, 230 °C, 220 °C, and 210 °C for 60 min, respectively. Sky-blue open diamonds show a temperature of 210 °C for 40 min. Although a dependence of the pretilt angle on temperature seems to be observed, the baking process failed, as evidenced by the unevenness in Figure 3b. The vertical axis in Figure 3 is the baking temperature and the horizontal axis is wt% SE-4811 in SE-150. By adjusting the baking temperature and time in accordance with the concentration, a linear relationship could be obtained for an RS of 73.0 mm. Although the precise details depend on the concentration, large quantities of SE-4811 cannot be baked at high temperatures. In contrast, large quantities of SE-150 cannot be baked at low temperatures. When either of the alignment films cannot be baked, a patchy pattern such as that shown in Figure 3 is observed. Additionally, because vertical alignment films are stronger than PIA, we believe that when blended, high pretilt angles can be achieved due to surface protection by horizontal alignment films. When blended at the optimal blending ratio, crossed-nicols observations reveal the fabrication of a clean cell with no visible patchiness.

3.1.4. The Pretilt Angle Mapping

For a blended alignment film consisting of a 6:4 blend of the horizontal alignment film PIA-X359-01X and the vertical alignment film PIA-X768-01X, we formed a cell using an RS of 35.3 mm and measured pretilt angles using an OPTIPRO-micro with a laser spot size of 3 μm. Figure 4 shows the results of pretilt angle mapping measurements for various measurement intervals and measurement ranges. We found that information on the peaks and valleys of finely measured pretilt angles could generally be obtained from plots even as the measurement intervals and measurement ranges were expanded; thus, for other samples, we measured pretilt angles at 160 μm intervals over an area of 10.2 mm^2. For the conditions described above, the gap between the peaks and valleys of the rubbing line was 3°, whereupon we conclude that we have fabricated a 60° ± 3° cell. The average and standard deviation of the pretilt angle mapping results obtained for other conditions are shown in Figure 5 and the mapping diagram in Figure 6. As has been reported in previous studies [20], even without the mixing of V-PI and H-PI and for a roughness of 500 nm, an optically averaged value emerges over a range of 3 μm. Although rubbing lines are clearly visible in some cells, overall, we obtain a fairly constant pretilt angle of about ±2.5°. For SE-based alignment films with the vertical alignment film component accounting for 67–90 wt% of the blend, the standard deviation was 15–4°, indicating that an appropriate baking temperature had not yet been reached. Considering the large error bars in Figure 5 from the mapping results in Figure 6, when the RS is 35.3 mm and the concentration of PIA-X768-01X is 80 wt% or higher, the error bars are considered to be large because the alignment film is repelled by the fine dust. In addition, when the RS is 35.3 mm and the concentration of PIA-X768-01X is 40 wt%, and when the RS is 73.0 mm and the concentration of SE-4811 is 66 or 82 wt%, it is unclear whether distortion occurred during cell assembly, but it is thought that the alignment film is distorted due to some effect.

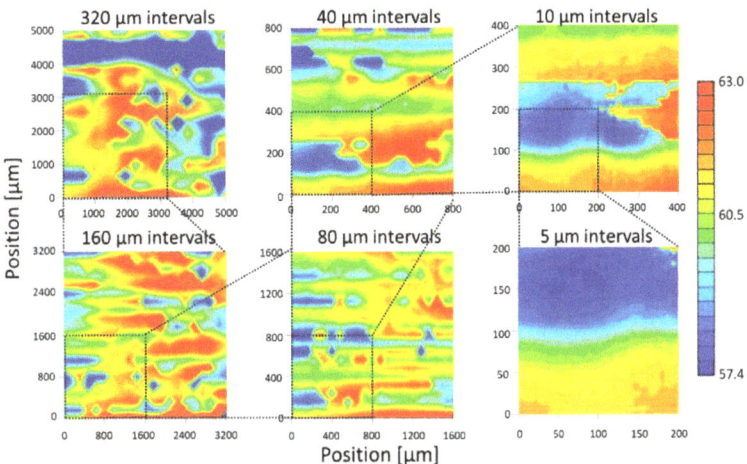

Figure 4. Pretilt angle mapping per measurement range.

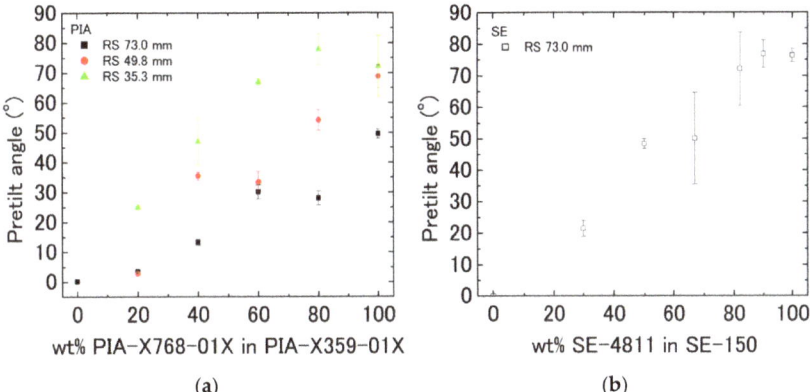

Figure 5. Average and standard deviation of pretilt angle within mapping range for (**a**) PIA-X768-01X and PIA-X359-01X and (**b**) SE-4811 and SE-150.

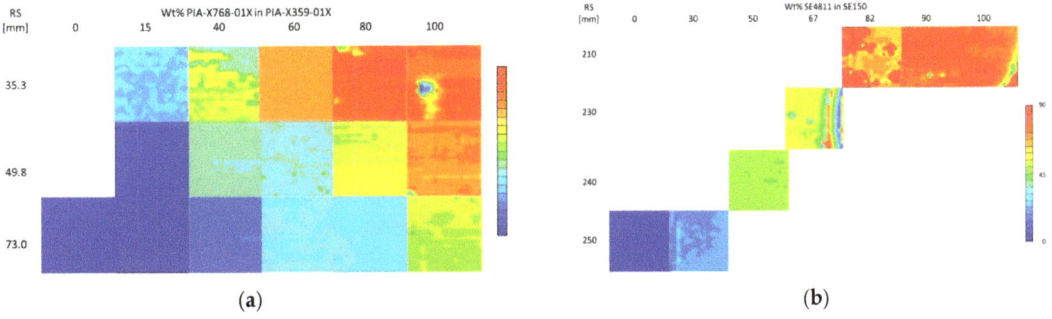

Figure 6. Pretilt angle mappings over an area of 10.2 mm^2 for (**a**) PIA-X768-01X and PIA-X359-01X and (**b**) SE-4811 and SE-150.

3.2. Viewing Angle for LCDs with High Pretilt Angle

Individual Devices

We injected ZLI-4792, to which 5 wt% of the dichroic dye NKX-4173 was added, to cells with pretilt angles of 22° and 48° for PIA-X359-01X/PIA-X768-01X blended alignment films and to cells with a thickness of 5 μm and a pretilt angle of 44° for SE-150/SE-4811 blended alignment films. Figure 7 shows photographs of the various cells, both untilted and tilted ±45° with respect to the vertical direction. We note that rubbing lines are somewhat visible. Next, to enhance the incidence angle dependence, we use a single polarizer to block light oscillating in directions not absorbed by the dye. Figure 8 shows measured data and simulation results for the incidence angle dependence of the transmittance of the various cells. As the pretilt angle increases, the peak transmittance angle shifts toward 0°; thus, by controlling the pretilt angle, we have successfully controlled the incidence angle. We attribute the slight shift in transmittance to a slight increase in pretilt angle due to the presence of the dye.

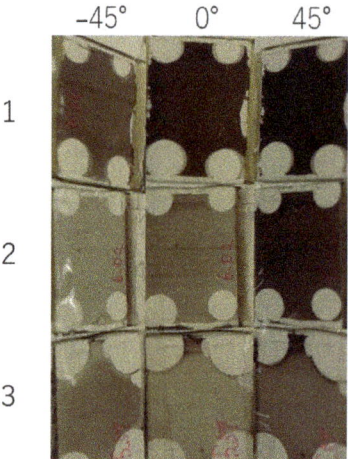

Figure 7. Photographs of GHLC cells with high pretilt angles, tilted by angles of −45°, 0°, and 45° about the vertical direction. The rows labeled 1, 2, and 3 in this figure correspond to cells with pretilt angles of 22°, 48°, and 44°, respectively.

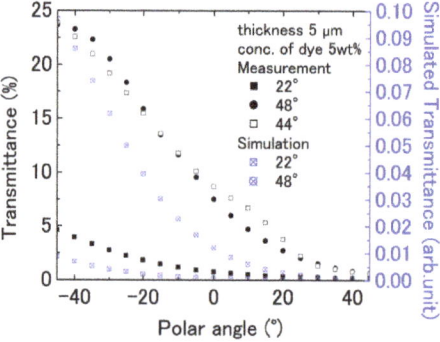

Figure 8. Angle dependence of transmittance for GHLC cells with high pretilt angles. Black squares, black circles, and open squares, respectively, indicate measured data for pretilt angles of 22°, 48°, and 44°. Square-enclosed crosses and circle-enclosed crosses, respectively, indicate simulation results for pretilt angles of 22° and 48°.

4. Conclusions

By blending horizontal and vertical alignment films, we succeeded in controlling the pretilt angle of an LC near the substrate interface. As a result, we could control the angle at which transmittance is maximized under no-voltage conditions. By adopting a guest/host LCD device with a pretilt angle of around 45°, we proposed a "louver" LCD with a transmittance peak near the 45° polar angle without the application of voltage. Using alignment films with identical baking temperatures allows pretilt angles to be controlled by varying the rubbing strength and the blending ratio for horizontal and vertical alignment films with no concern for baking temperature. In contrast, using alignment films with different baking temperatures requires that the temperature be adjusted depending on the blending ratio but still allows pretilt angles to be controlled with a range of roughly ±3°. Differences in the ability of vertical alignment films to withstand rubbing lead to regions in which the film is highly concentrated, reducing pretilt angles and causing rubbing lines to be prominently visible. Variations on the order of ±3° suggest that these devices may not be usable for fine-grained display systems but should be acceptable as devices with an incidence angle dependence. Indeed, for devices combining dichroic dyes with polarizers, we expect that rubbing lines and variations in pretilt angle should not be major sources of concern. By controlling the pretilt angle, we have succeeded in controlling the peak transmittance angle and fabricating LC devices with an incidence angle dependence. We expect that the device will be developed into viewing-angle-dependent devices for building material windows and car windows in the future.

Author Contributions: Conceptualization, M.I., E.F., M.A. and K.T.; methodology, M.I., E.F., M.A. and K.T.; Formal analysis, M.I.; investigation, M.I., H.H., Y.N. and H.T.; resources, M.A. and K.T.; data curation, M.I.; writing—original draft preparation, M.I.; writing—review and editing, M.I., E.F., M.A. and K.T.; visualization, M.I.; supervision, M.I. and K.T.; project administration, M.I.; funding acquisition, M.I. All authors have read and agreed to the published version of the manuscript.

Funding: This research received no external funding.

Institutional Review Board Statement: Not applicable.

Informed Consent Statement: Not applicable.

Data Availability Statement: Not applicable.

Acknowledgments: The authors are thankful to Merck & Company Incorporated for providing the LC materials, to Hayashibara Limited for providing the dichroic dyes, and to Nissan Chemical and Chisso Petrochemical Corporation for providing the polyimide materials.

Conflicts of Interest: The authors declare no conflict of interest.

References

1. Yamaguchi, R.; Inoue, K.; Oikawa, Y.; Takasu, T. Electro-Optical Property in Hybrid Aligned Reverse mode Cell Using Liquid Crystals with Positive and Negative Dielectric Constant Anisotropies. *J. Photopolym. Sci. Technol.* **2017**, *30*, 463–466. [CrossRef]
2. Yamaguchi, R.; Waki, Y.; Sato, S. Wide Viewing Angle Properties in Nematic Liquid Crystal/UV Curable Liquid Crystal Composite Films with Some Aligned-Modes. *J. Photopolym. Sci. Technol.* **1997**, *10*, 19–24. [CrossRef]
3. Yamaguchi, R.; Ushizaki, R. Louver Function in Hybrid Aligned Reverse Mode Using Dual Frequency Liquid Crystal. *J. Photopolym. Sci. Technol.* **2019**, *32*, 545–548. [CrossRef]
4. Kasajima, Y.; Kato, T.; Kubono, A.; Tasaka, S.; Akiyama, R. Wide Viewing Angle of Rubbing-Free Hybrid Twisted Nematic Liquid Crystal Displays. *Jpn. J. Appl. Phys.* **2008**, *47*, 7941–7942. [CrossRef]
5. Takatoh, K.; Ito, M.; Saito, S.; Takagi, Y. Optical Filter with Large Angular Dependence of Transmittance Using Liquid Crystal Devices. *Crystals* **2021**, *11*, 1199. [CrossRef]
6. Geary, J.M.; Goodby, J.W.; Kmetz, A.R.; Patel, J.S. The mechanism of polymer alignment of liquid-crystal materials. *J. Appl. Phys.* **1987**, *62*, 4100–4108. [CrossRef]
7. Hasebe, H.; Takatsu, H.; Iimura, Y.; Kobayashi, S. Effect of Polymer Network Made of Liquid Crystalline Diacrylate on Characteristics of Liquid Crystal Display Device. *Jpn. J. Appl. Phys.* **1994**, *33*, 6245. [CrossRef]
8. Chang, K.-H.; Joshi, V.; Chien, L.-C. Fast-switching chiral nematic liquid-crystal mode with polymer-sustained twisted vertical alignment. *Phys. Rev. E* **2017**, *95*, 042701. [CrossRef]

9. Mundinger, S.; Gnauck, S.; Tong, Q.; Weegels, L.; Terfort, A. Multi-reactive mesogen system for polymer-stabilised vertical alignment liquid crystal displays. *Liq. Cryst.* **2021**, *49*, 209–216. [CrossRef]
10. Fujisawa, T.; Jang, K.; Kodera, F.; Gushiken, M.; Sudou, G.; Hasebe, H.; Takatsu, H. 31.4: Properties of Nano-Phase-Separated Liquid Crystals (NPS LCs) with Fast Response. *SID Symp. Dig. Tech. Pap.* **2015**, *46*, 458–461. [CrossRef]
11. Fujisawa, T.; Jang, K.; Kodera, F.; Hasebe, H.; Takatsu, H. 27-2: Novel Photo-Polymer Stabilization of Nano-Phase-Separated LCs with Fast Response. *SID Symp. Dig. Tech. Pap.* **2017**, *48*, 381–384. [CrossRef]
12. Wang, R.; Atherton, T.J.; Zhu, M.; Petschek, R.G.; Rosenblatt, C. Naturally occurring reverse tilt domains in a high-pretilt alignment nematic liquid crystal. *Phys. Rev. E* **2007**, *76*, 021702. [CrossRef] [PubMed]
13. Vaughn, K.E.; Sousa, M.; Kang, D.; Rosenblatt, C. Continuous control of liquid crystal pretilt angle from homeotropic to planar. *Appl. Phys. Lett.* **2007**, *90*, 194102. [CrossRef]
14. Wu, W.-Y.; Wang, C.-C.; Fuh, A.Y.-G. Controlling pre-tilt angles of liquid crystal using mixed polyimide alignment layer. *Opt. Express* **2008**, *16*, 17131–17137. [CrossRef] [PubMed]
15. Holmes, C.J.; Taphouse, T.S.; Sambles, J.R. Characterizing Two Methods for Achieving Intermediate Surface Pretilt. *Mol. Cryst. Liq. Cryst.* **2012**, *553*, 81–89. [CrossRef]
16. Takatoh, K.; Akimoto, M.; Kaneko, H.; Kawashima, K.; Kobayashi, S. Molecular arrangement for twisted nematic liquid crystal displays having liquid crystalline materials with opposite chiral structures (reverse twisted nematic liquid crystal displays). *J. Appl. Phys.* **2009**, *106*, 064514. [CrossRef]
17. Kim, T.; Ju, C.; Kang, H. LC alignment behaviors on polystyrene blend alignment films. *Mol. Cryst. Liq. Cryst.* **2018**, *664*, 38–45. [CrossRef]
18. Kim, J.B.; Kim, K.C.; Ahn, H.J.; Hwang, B.H.; Jo, S.J.; Kim, C.S.; Baik, H.K.; Choi, C.J.; Jo, M.K.; Kim, Y.S.; et al. No bias pi cell using a dual alignment layer with an intermediate pretilt angle. *Appl. Phys. Lett.* **2007**, *91*, 023507. [CrossRef]
19. Lee, Y.-J.; Gwag, J.S.; Kim, Y.-K.; Jo, S.I.; Kang, S.-G.; Park, Y.R.; Kim, J.-H. Control of liquid crystal pretilt angle by anchoring competition of the stacked alignment layers. *Appl. Phys. Lett.* **2009**, *94*, 041113. [CrossRef]
20. Yeung, F.S.; Ho, J.Y.; Li, Y.W.; Xie, F.C.; Tsui, O.K.; Sheng, P.; Kwok, H.S. Variable liquid crystal pretilt angles by nanostructured surfaces. *Appl. Phys. Lett.* **2006**, *88*, 051910. [CrossRef]
21. Ishihara, S.; Mizusaki, M. Alignment control technology of liquid crystal molecules. *J. Soc. Inf. Disp.* **2019**, *28*, 44–74. [CrossRef]
22. Scheffer, T.J.; Nehring, J. Accurate determination of liquid-crystal tilt bias angles. *J. Appl. Phys.* **1977**, *48*, 1783–1792. [CrossRef]
23. Pauluth, D.; Tarumi, K. Advanced liquid crystals for television. *J. Mater. Chem.* **2004**, *14*, 1219–1227. [CrossRef]
24. Fukuro, H.; Kobayashi, S. Newly Synthesized Polyimide for Aligning Nematic Liquid Crystals Accompanying High Pretilt Angles. *Mol. Cryst. Liq. Cryst. Inc. Nonlinear Opt.* **1988**, *163*, 157–162. [CrossRef]

Disclaimer/Publisher's Note: The statements, opinions and data contained in all publications are solely those of the individual author(s) and contributor(s) and not of MDPI and/or the editor(s). MDPI and/or the editor(s) disclaim responsibility for any injury to people or property resulting from any ideas, methods, instructions or products referred to in the content.

Article

Multichromic Behavior of Liquid Crystalline Composite Polymeric Films

Mizuho Kondo [1,*], Satoka Yanai [1], Syouma Shirata [1], Takeshi Kakibe [2], Jun-ichi Nishida [1] and Nobuhiro Kawatsuki [1]

[1] Department of Applied Chemistry, Graduate School of Engineering, University of Hyogo, 2167 Shosha, Himeji 671-2280, Japan; et21c002@steng.u-hyogo.ac.jp (S.Y.); r4.sshirata@gmail.com (S.S.); jnishida@eng.u-hyogo.ac.jp (J.-i.N.); kawatuki@eng.u-hyogo.ac.jp (N.K.)

[2] Department of Chemical Engineering, Graduate School of Engineering, University of Hyogo, 2167 Shosha, Himeji 671-2280, Japan; kakibet@eng.u-hyogo.ac.jp

* Correspondence: mizuho-k@eng.u-hyogo.ac.jp; Tel.: +81-79-267-4014

Abstract: In this study, we describe the synthesis of a cholesterol-linked cyanostilobazole salt dye and the tuning of its luminescence by physical stimuli such as electricity and grinding. The dyes exhibited liquid-crystalline properties at temperatures above 170 °C. Some of the solutions were transformed into orange luminescent gels upon the addition of poor solvents. When the solvent was evaporated, the resulting solid xerogel exhibited mechanochromism, its color changed, and its luminescent color changed from orange to red. Furthermore, we investigated the construction of functional gels (mechanochromic gels) that can respond to two stimuli, damage detection by abrasive responsiveness, and electrical response using ionic liquid complexes of polymers as dispersing media. This study provides a new strategy for tuning and switching luminescence using non-chemical stimuli in a single-component system using aggregation.

Keywords: mechanofluorochromic dye; liquid crystal; gel; electrochromism

Citation: Kondo, M.; Yanai, S.; Shirata, S.; Kakibe, T.; Nishida, J.-i.; Kawatsuki, N. Multichromic Behavior of Liquid Crystalline Composite Polymeric Films. *Crystals* **2023**, *13*, 786. https://doi.org/10.3390/cryst13050786

Academic Editor: Francesco Simoni

Received: 28 April 2023
Revised: 3 May 2023
Accepted: 4 May 2023
Published: 9 May 2023

Copyright: © 2023 by the authors. Licensee MDPI, Basel, Switzerland. This article is an open access article distributed under the terms and conditions of the Creative Commons Attribution (CC BY) license (https://creativecommons.org/licenses/by/4.0/).

1. Introduction

Low-molecular-weight gels are materials with a continuous structure of macroscopic dimensions composed of a low concentration of gelator and solvent, and exhibit rheological behavior similar to that of solids. Gels formed by low-molecular-weight gelators have been put to practical use as aggregating agents, taking advantage of their ability to switch between flowable sol and solid-state gel upon exposure to external stimuli. They are also expected to be applied in a wide range of fields, such as drug delivery [1], sensors, and self-healing materials [2], and are being intensively studied. Recently, aggregation-induced emission (AIE) gelators, in which a luminescent moiety is introduced into low-molecular-weight gelators and the luminescence behavior changes significantly during the gel-to-sol and sol-to-gel phase transitions, have attracted attention from both theoretical and practical perspectives [3,4]. AIE gels have the potential to be excellent stimulus-responsive luminescent materials because their luminescence behavior changes in tandem with the aggregation and disassembly of supramolecular networks induced by various external stimuli. In particular, the mechano-responsive property, in which the luminescence behavior is changed by grinding, allows macroscopic mechanical stimuli to affect the optical properties of molecules and is expected to be applied to displacement and force sensors with excellent spatial resolution [5,6]. In recent years, many mechanochromic gels utilizing the intermolecular interactions of chitosan [7] and cholesterol [4], as well as mechanosensitive gels utilizing the cholesteric phase of liquid crystals [8] or the change in selective reflection of photonic crystals [9], have been reported. Small-molecule gels are generally produced by introducing long hydrophobic tails, such as cholesterol, alkyl chains, and alkoxy chains [10]. Many small-molecule organogelators have been investigated, among which cholesterol has been extensively studied because of its ability to form stable gels through the synergistic effect of cooperative noncovalent bonding. However, cholesterol has a relatively rigid fused ring structure and can be used as a substituent for the development of liquid crystallinity. Furthermore, its gelation ability is expected to be enhanced without impairing liquid crystallinity. In dyes with the introduced liquid crystalline properties, functional mechanochromic (MC) behaviors can be realized, such as phase-dependent color control [11], multicolor luminescence [12,13], metastable color stability control [14], and orientation control [15,16]. Therefore, several research groups have investigated the combination of liquid crystalline properties and aggregation-induced luminescence. By introducing trans, a mesogen widely used to produce liquid crystals, more stable and aggregation-responsive polysynthetic dyes with excellent aggregation properties can be obtained. In a previous study, we synthesized a compound consisting of cholesterol, a gel-forming functional group, a cyanostilbene derivative, and a friability-responsive moiety linked to phenylacetylene. Cyanostilbene has a large dipole moment and twisted geometry along the central ethylene unit. It also tends to form stacked molecular sheets through multiple C–H—N and C–H—O hydrogen bonds. As these structures can be modified by external stimuli, they are frequently used as functional groups to express MC luminescence. Cyanostilbene derivatives also have donor–acceptor structures and tend to form antiparallel molecular pairs. The formation of molecular pairs leads to the formation of bifunctional cholesterol at the ends of the complexes, which are expected to form gels when linked one-dimensionally via intermolecular interactions [4,17]. In addition, we reported that the spin-coated film formed from the sol state exhibits a friability response, and that, by utilizing its liquid crystalline nature, it exhibits polarized luminescence when oriented parallel to the friability direction, enabling the recognition of the friability direction [18].

In the gel state, various functional liquids can be incorporated into the dispersing medium to form flexible functional composites and combining properties that are difficult to achieve using a single compound. In this study, we focused on ionic liquids (ILs) as a dispersing medium for functionalities because IL-composite polymer composite gels can be used as conductive materials and are expected to be applied for damage detection in conductive materials by providing a mechanical stimulus response. In addition, if the composite is electrically responsive (electrochromism (EC)) to ILs, functional gels (mechanochromic gels) that respond to both electrical and mechanical stimuli can be constructed. The addition of an EC response simplifies the control of coloration and responsiveness, enables flexible control of the physical properties, and can be applied to the detection of breakage and short circuiting of the electrolyte inside a battery. Although some studies have introduced MC dyes into films of conductors to enable the detection of external forces [19], no study has directly measured the load on a conductor. This is because the change in the MC dye manifests itself in crystalline and solid systems, whereas the electrical response manifests itself in solution systems, making it difficult to achieve compatibility. In this study, we attempted to achieve stimulus-responsiveness using the dye itself as a physical gel.

2. Results and Discussion

2.1. General Property of the Luminophore

Figure 1 shows the chemical structures of the luminophores used in this study. The synthesis and evaluation methods of the compounds are specifically described in the experimental section. The method of evaluation of the mechanoresponse is a method employed by many research groups. In a previous study, phenylacetylene was introduced at the end of the conjugated backbone; however, thiophene, which has a higher HOMO and smaller energy gap, was introduced in this study. Because cyanostyrobazole linked to thiophene exhibits a friability response even in the protonated state [17], friability is also expected in quaternized compounds. Therefore, to facilitate the expression of the electrical response, we changed the molecular end of the salt to a quaternized pyridine to make it a donor–acceptor salt. The optimized structures were determined using density functional

Article

Multichromic Behavior of Liquid Crystalline Composite Polymeric Films

Mizuho Kondo [1,*], Satoka Yanai [1], Syouma Shirata [1], Takeshi Kakibe [2], Jun-ichi Nishida [1] and Nobuhiro Kawatsuki [1]

[1] Department of Applied Chemistry, Graduate School of Engineering, University of Hyogo, 2167 Shosha, Himeji 671-2280, Japan; et21c002@steng.u-hyogo.ac.jp (S.Y.); r4.sshirata@gmail.com (S.S.); jnishida@eng.u-hyogo.ac.jp (J.-i.N.); kawatuki@eng.u-hyogo.ac.jp (N.K.)

[2] Department of Chemical Engineering, Graduate School of Engineering, University of Hyogo, 2167 Shosha, Himeji 671-2280, Japan; kakibet@eng.u-hyogo.ac.jp

* Correspondence: mizuho-k@eng.u-hyogo.ac.jp; Tel.: +81-79-267-4014

Abstract: In this study, we describe the synthesis of a cholesterol-linked cyanostilobazole salt dye and the tuning of its luminescence by physical stimuli such as electricity and grinding. The dyes exhibited liquid-crystalline properties at temperatures above 170 °C. Some of the solutions were transformed into orange luminescent gels upon the addition of poor solvents. When the solvent was evaporated, the resulting solid xerogel exhibited mechanochromism, its color changed, and its luminescent color changed from orange to red. Furthermore, we investigated the construction of functional gels (mechanochromic gels) that can respond to two stimuli, damage detection by abrasive responsiveness, and electrical response using ionic liquid complexes of polymers as dispersing media. This study provides a new strategy for tuning and switching luminescence using non-chemical stimuli in a single-component system using aggregation.

Keywords: mechanofluorochromic dye; liquid crystal; gel; electrochromism

1. Introduction

Low-molecular-weight gels are materials with a continuous structure of macroscopic dimensions composed of a low concentration of gelator and solvent, and exhibit rheological behavior similar to that of solids. Gels formed by low-molecular-weight gelators have been put to practical use as aggregating agents, taking advantage of their ability to switch between flowable sol and solid-state gel upon exposure to external stimuli. They are also expected to be applied in a wide range of fields, such as drug delivery [1], sensors, and self-healing materials [2], and are being intensively studied. Recently, aggregation-induced emission (AIE) gelators, in which a luminescent moiety is introduced into low-molecular-weight gelators and the luminescence behavior changes significantly during the gel-to-sol and sol-to-gel phase transitions, have attracted attention from both theoretical and practical perspectives [3,4]. AIE gels have the potential to be excellent stimulus-responsive luminescent materials because their luminescence behavior changes in tandem with the aggregation and disassembly of supramolecular networks induced by various external stimuli. In particular, the mechano-responsive property, in which the luminescence behavior is changed by grinding, allows macroscopic mechanical stimuli to affect the optical properties of molecules and is expected to be applied to displacement and force sensors with excellent spatial resolution [5,6]. In recent years, many mechanochromic gels utilizing the intermolecular interactions of chitosan [7] and cholesterol [4], as well as mechanosensitive gels utilizing the cholesteric phase of liquid crystals [8] or the change in selective reflection of photonic crystals [9], have been reported. Small-molecule gels are generally produced by introducing long hydrophobic tails, such as cholesterol, alkyl chains, and alkoxy chains [10]. Many small-molecule organogelators have been investigated, among

Citation: Kondo, M.; Yanai, S.; Shirata, S.; Kakibe, T.; Nishida, J.-i.; Kawatsuki, N. Multichromic Behavior of Liquid Crystalline Composite Polymeric Films. *Crystals* **2023**, *13*, 786. https://doi.org/10.3390/cryst13050786

Academic Editor: Francesco Simoni

Received: 28 April 2023
Revised: 3 May 2023
Accepted: 4 May 2023
Published: 9 May 2023

Copyright: © 2023 by the authors. Licensee MDPI, Basel, Switzerland. This article is an open access article distributed under the terms and conditions of the Creative Commons Attribution (CC BY) license (https://creativecommons.org/licenses/by/4.0/).

which cholesterol has been extensively studied because of its ability to form stable gels through the synergistic effect of cooperative noncovalent bonding. However, cholesterol has a relatively rigid fused ring structure and can be used as a substituent for the development of liquid crystallinity. Furthermore, its gelation ability is expected to be enhanced without impairing liquid crystallinity. In dyes with the introduced liquid crystalline properties, functional mechanochromic (MC) behaviors can be realized, such as phase-dependent color control [11], multicolor luminescence [12,13], metastable color stability control [14], and orientation control [15,16]. Therefore, several research groups have investigated the combination of liquid crystalline properties and aggregation-induced luminescence. By introducing trans, a mesogen widely used to produce liquid crystals, more stable and aggregation-responsive polysynthetic dyes with excellent aggregation properties can be obtained. In a previous study, we synthesized a compound consisting of cholesterol, a gel-forming functional group, a cyanostilbene derivative, and a friability-responsive moiety linked to phenylacetylene. Cyanostilbene has a large dipole moment and twisted geometry along the central ethylene unit. It also tends to form stacked molecular sheets through multiple C–H—N and C–H—O hydrogen bonds. As these structures can be modified by external stimuli, they are frequently used as functional groups to express MC luminescence. Cyanostilbene derivatives also have donor–acceptor structures and tend to form antiparallel molecular pairs. The formation of molecular pairs leads to the formation of bifunctional cholesterol at the ends of the complexes, which are expected to form gels when linked one-dimensionally via intermolecular interactions [4,17]. In addition, we reported that the spin-coated film formed from the sol state exhibits a friability response, and that, by utilizing its liquid crystalline nature, it exhibits polarized luminescence when oriented parallel to the friability direction, enabling the recognition of the friability direction [18].

In the gel state, various functional liquids can be incorporated into the dispersing medium to form flexible functional composites and combining properties that are difficult to achieve using a single compound. In this study, we focused on ionic liquids (ILs) as a dispersing medium for functionalities because IL-composite polymer composite gels can be used as conductive materials and are expected to be applied for damage detection in conductive materials by providing a mechanical stimulus response. In addition, if the composite is electrically responsive (electrochromism (EC)) to ILs, functional gels (mechanochromic gels) that respond to both electrical and mechanical stimuli can be constructed. The addition of an EC response simplifies the control of coloration and responsiveness, enables flexible control of the physical properties, and can be applied to the detection of breakage and short circuiting of the electrolyte inside a battery. Although some studies have introduced MC dyes into films of conductors to enable the detection of external forces [19], no study has directly measured the load on a conductor. This is because the change in the MC dye manifests itself in crystalline and solid systems, whereas the electrical response manifests itself in solution systems, making it difficult to achieve compatibility. In this study, we attempted to achieve stimulus-responsiveness using the dye itself as a physical gel.

2. Results and Discussion

2.1. General Property of the Luminophore

Figure 1 shows the chemical structures of the luminophores used in this study. The synthesis and evaluation methods of the compounds are specifically described in the experimental section. The method of evaluation of the mechanoresponse is a method employed by many research groups. In a previous study, phenylacetylene was introduced at the end of the conjugated backbone; however, thiophene, which has a higher HOMO and smaller energy gap, was introduced in this study. Because cyanostyrobazole linked to thiophene exhibits a friability response even in the protonated state [17], friability is also expected in quaternized compounds. Therefore, to facilitate the expression of the electrical response, we changed the molecular end of the salt to a quaternized pyridine to make it a donor–acceptor salt. The optimized structures were determined using density functional theory (DFT) calculations. Owing to limitations of the computational environment, the thiophene end was assumed to be a methyl group. Frontier orbital plots of the HOMO and LUMO of the optimized structure are shown in Figure 1b,c, respectively; the LUMO was mainly distributed in the acceptor of the cyanostilbene unit, while the HOMO was mainly located in the alkylthiophene unit. The energy levels of HOMO and LUMO are estimated to be 8.05 and 6.13 eV, respectively, with an energy gap between these levels of 1.92 eV. The energy-minimized structures exhibited twisted conformations and improved planarity of the cyanostilobazole moiety compared with the compounds before cholesterol introduction. These electronic structures were similar to those of the protonated cyanostyrobazole derivatives; cholesterol was predicted to be present in a bent conformation at right angles from the conjugated moiety. This is because, in the previous structure, cholesterol was linked to the quaternized cyanostyrobazole via a bromoacetate ester, whereas in the present structure, cholesterol was linked as a benzoate ester, and the linearity was reduced [18].

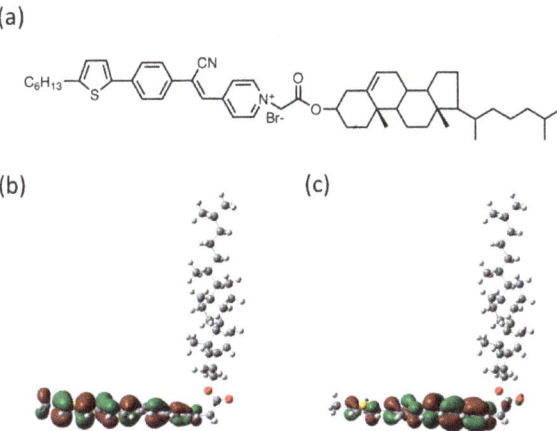

Figure 1. (a) Chemical structures and (b) HOMO and (c) LUMO of the luminophore used in this study. Red and green lobes represent positive and negative coefficients for molecular orbitals, respectively, and the size of each lobe is proportional to the magnitude of these coefficients.

These compounds were dissolved in various solvents, and their absorption and emission spectra were measured. The dyes were dissolved in common solvents, including toluene, tetrahydrofuran (THF), dimethyl sulfoxide (DMSO), and dichloromethane (DCM), which have increasing polarity in the order of toluene < THF < DCM < DMSO. An overview of the changes in the absorption spectra of the dyes dissolved in various solvents is shown in Figure 2a. The absorption shifted to longer wavelengths in more polar solvents. The differences in the luminescence behavior was more pronounced, with a significant red-shift in the photoluminescence (PL) spectral peak when dissolved in DMSO or DCM (Figure 2b).

The effect of the solvent polarity on the absorption spectrum was estimated using time-dependent TD-DFT calculations, and the results are shown in Figure 3. To simplify the calculations, a model compound with a methyl group at the styrobazole end, independent of the conjugation, was used. The λ_{abs} showed red shifts in the order DMSO < DCM < THF < toluene, which is opposite to the experimental results. All of these absorptions are HOMO-LUMO transitions. The reason for this discrepancy may be that the interaction with the solvent is not simple, and the contribution of Br as a counter ion was neglected in the calculations.

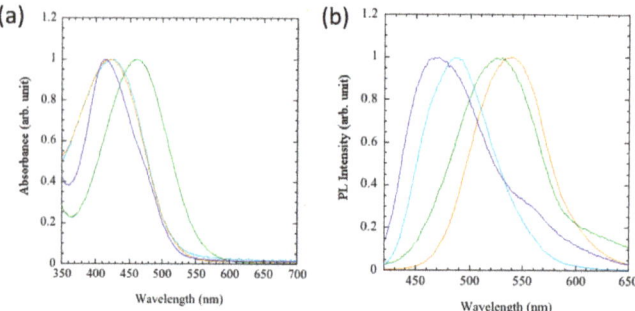

Figure 2. (a) Absorption and (b) photoluminescence (PL) spectra of the luminophore (1.0×10^{-4} mol/L) dissolved in toluene, (blue) dichloromethane, (DCM, green) tetrahydrofuran (THF, blue), and dimethyl sulfoxide (DMSO, orange).

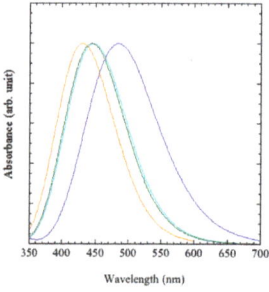

Figure 3. TD-DFT calculated absorption spectra of the luminophore dissolved in toluene (blue), DCM (green), THF (blue), and DMSO (orange).

2.2. Mechanoresponsive Behavior

The mechanical responses of the compound powders were evaluated. Figure 4a,b show the PL spectra and luminescence photographs before and after grinding, respectively. Figure 4c shows the change in the luminescence decay profile before and after grinding. Before grinding, the powder showed orange emission with a maximum emission wavelength of 637 nm and a fluorescence lifetime of 7.22 ns; after grinding in an agate mortar, the luminescence changed to red, with a maximum emission wavelength of 654 nm and a fluorescence lifetime of 7.44 ns, which did not change after additional grinding. The luminescence quantum yield changed from 1.8% to 1.7%. Unlike the previously gelated material, no long-lived luminescent components were detected, indicating excimer formation. Therefore, it can be inferred that excimer formation is not essential for a series of derivatives capable of gelation. In addition, the luminescence intensity did not change drastically with grinding, as in the case of the AIE-mediated dye, suggesting that the luminescence intensity changed via a mechanism different from that of AIE.

Scanning electron microscopy (SEM) observation of the powder was also performed on the powders. Fibrous structures were observed on the surface of the powder before grinding but not after grinding. Figure 5b shows the powder X-ray diffraction (XRD) results before and after the grinding of the powder. Some peaks were observed in the powder that disappeared after grinding. Compared to previous compounds, the peaks were broad and low in intensity. Therefore, the crystallinity was low and the powder was amorphous after grinding.

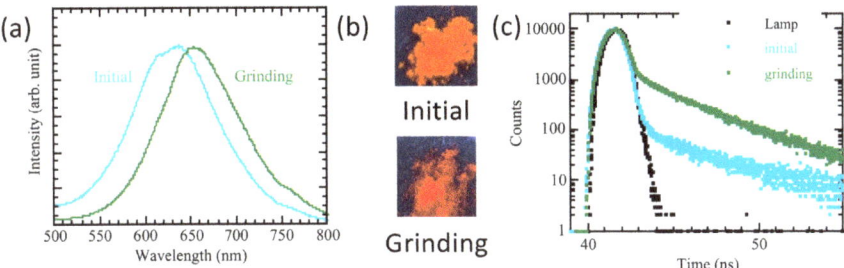

Figure 4. Change in (**a**) PL spectrum, (**b**) photoluminescent color, (**c**) and fluorescent decay profile of the luminophore before and after mechanical grinding.

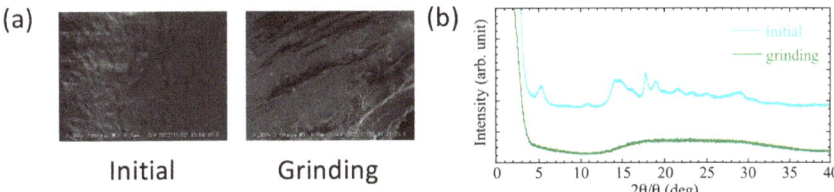

Figure 5. (**a**) Scanning electron microscopy images and (**b**) X-ray diffraction patterns before and after mechanical grinding. The scalebar in (**a**): 2 μm.

When the ground powder was heated, it formed a birefringent fluid phase and solidified when cooled to room temperature. The maximum emission wavelength remained unchanged at 637 nm during the series of operations, and the emission color did not change when the film was ground again. Differential scanning calorimetry (DSC) measurements of the dye showed that the slope of the thermogram changed at temperatures above 150 °C, suggesting that a change in the solid state was induced (Figure 6). The thermal properties disappeared in the second scan when the temperature was scanned up to 230 °C.

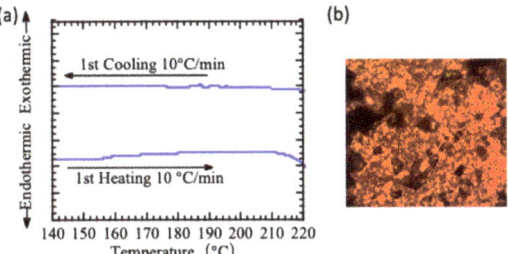

Figure 6. (**a**) Differential scanning calorimetry thermogram of the luminophore and (**b**) polarized optical microscopy image of the luminophore at 170 °C.

The fluid phase was presumed to be the liquid crystalline phase; however, unlike in a previous report, no optical structure derived from the cholesteric phase could be observed. This may be due to the fact that the cholesteric and rigid luminescent sites are not linearly connected and the liquid crystallinity is not stable.

While the compound did not gel in a single solvent, it gelled at a concentration of 14.3 g/L in a 6/1 (v/v) mixture of chloroform as a good solvent and hexane as a poor solvent (Figure 7). The gel showed fluidity when heated to 40 °C and reversible gelation when cooled to 4 °C.

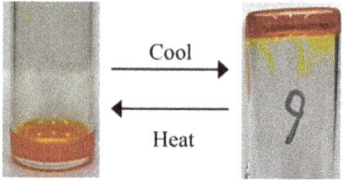

Figure 7. Inversion test of the luminophore dissolved in 6/1 (*v*/*v*) mixture of chloroform and hexane.

The material that changed color upon gelation retained its luminescent color even after the solvent was removed from the xerogel and exhibited a MC response. The changes in the PL spectra are shown in Figure 8a,b. The XRD patterns showed no diffraction peaks before and after grinding (Figure 8c). From SEM observations, the surface of the ground material showed a pattern similar to that of a fibrous structure before grinding but not after grinding (Figure 8d).

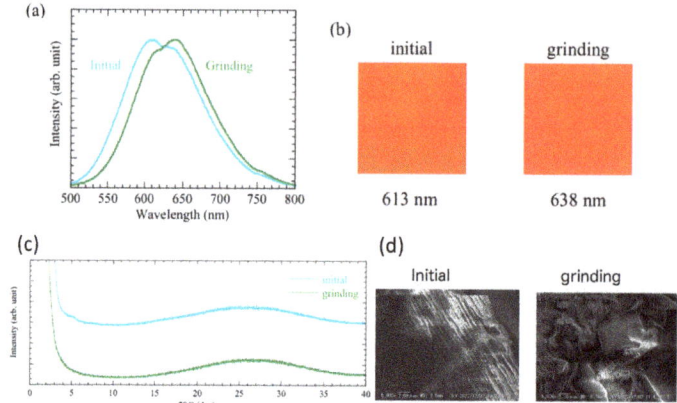

Figure 8. Mechano-induced changes in (**a**) PL spectrum, (**b**) photoluminescent color, (**c**) X-ray diffractogram, and (**d**) SEM image of the luminophore film prepared from the xerogel. The scalebar in (**d**): 2 μm.

These results suggest that the color change is not due to the crystalline structure but rather to the one-dimensional structure of the solid and gel interiors and that less rigorous intermolecular interactions are responsible for the color change.

2.3. Electroresponsive Behavior

Materials that simultaneously exhibit AIE and EC have been reported by Sun et al. [20]. In addition, as mentioned in the introduction, some papers have reported that AIE-responsive materials connected to cholesterol exhibit MC. Taking these reports into consideration, we can expect to find cholesterol-based MC materials that exhibit EC. However, since EC is mainly a response in liquids and MC is mainly a response in solids, there have been no reports of systems in which EC and MC can be simultaneously expressed. On the other hand, there have been several reports of materials that exhibit EC in a gel composite [21–23], and therefore, it is expected that EC and MC can be expressed in the same system by utilizing a material that exhibits MC in a gel state. Accordingly, for the AIE system for cholesterol, we attempted EC with a material with a lower band gap based on the previous material.

The electrical response of the luminophore was evaluated by cyclic voltammetry measurements in DCM in the presence of tetrabutylammonium hexafluorophosphate (n-Bu$_4$NClO$_4$) (0.1 M) with ferrocene Fc/Fc$^+$ as an internal standard, showing only an ambiguous one-electron oxidation at 0.18 V, indicating an irreversible electrical response (Figure 9). Based on these electrical responses, the HOMO and LUMO levels of the compound were estimated using the following equations:

$$\text{HOMO} = -(E_{ox} + 4.71)$$

$$\text{LUMO} = -(E_{red} + 4.71)$$

where E_{ox} and E_{red} are the oxidation and reduction voltages, respectively; the HOMO and LUMO are estimated to be −6.01 and −3.91 eV, respectively, and the band gap between these levels is 2.10 eV, which is close to the values estimated from the DFT calculations.

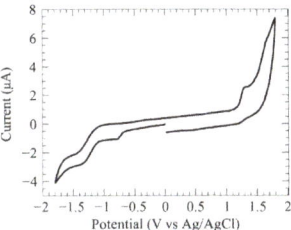

Figure 9. Cyclic voltammograms of the luminophore. Electrolytic medium: 0.1 M NBu$_4$PF$_6$-dichloromethane.

Compounding the luminophore with ILs was also investigated. The typical IL 1-ethyl-3-methylimidazolium/bis(trifluoromethanesulfonyl)imide (EMI/TFSI) was mixed with the dye to investigate its solvent-induced gelation ability and MC response of the gel films. A solution of 11.1 g/L in a 6/1 (v/v) chloroform/hexane mixture showed a sol–gel transition at 40 °C and a reversible sol–gel phase transition upon heating and cooling. The electrochemical response was evaluated using the experimental set-up shown in Figure 10a. Figure 10b shows the absorption spectra and color changes when a voltage of −2 V was applied to the poly (sodium 4-stylene sulfonate) side of the anode. The absorption at 450 nm decreased, and the color of the entire cell became lighter. The color did not change when the cathode and anode were switched. Compared to the data reported in a previous paper [18], the maximum absorption wavelength is shorter than that of non-quaternized oligothiophene. Therefore, the luminophores may have been decomposed, as well as cleavage of the π-conjugated structure by the reduction potential, resulting in an irreversible electrical response. However, the compound complexed with the ILs showed a low intensity, and no friability response was observed in the spin-coated thin films.

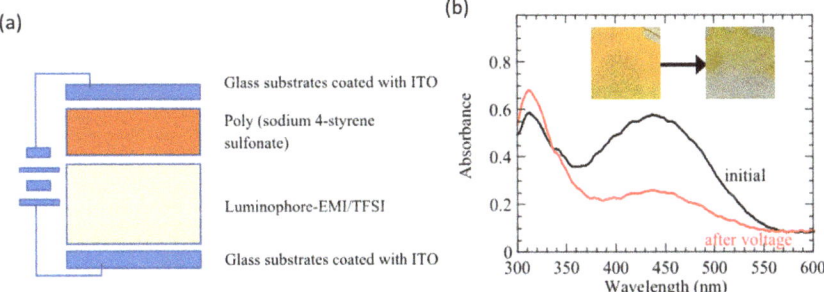

Figure 10. (**a**) Schematic of the experimental setup and (**b**) change in absorption spectrum of the luminophore dissolved in EMI/TFSI. The insets in (**b**) are photographs of the sample before (top) and after (bottom) applying −2 V for 3 min.

To improve the mechanical strength of the composite, a ternary composite of polymer, IL, and dye was investigated. Shea et al. previously reported that an IL of an imidazole derivative mixed with polymethylmethacrylate (PMMA) in a 1:1 weight ratio forms a flexible and conductive gel-like composite that can be used as an electrode [24]. The 1:1 mixture of EMI/TFSI and PMMA and the luminophore showed gelation in a 1/6 (v/v) chloroform/hexane mixture and reversible sol–gel phase transition upon heating and cooling at 40 °C; both sol–gels showed orange luminescence under UV light irradiation. The composite solution was cast onto a glass substrate, and its response to grinding was evaluated. Figure 11a shows the changes in the PL spectrum before and after grinding. Before grinding, the powder exhibited a PL band at 637 nm, which shifted to 654 nm after grinding.

Figure 11. Change in (**a**) the PL spectrum upon mechanical grinding and (**b**) the absorption spectrum upon applying −3 V for 3 min of the luminophore-EMI/TFSI-polymer composite. The inset in (**a**,**b**) are photographs of the composite under UV light and ambient light, respectively.

Furthermore, the electrical response was examined using the experimental system shown in Figure 11b. An irreversible color change was observed in the polymer-incorporated composite, as well as in the IL composite. Therefore, a system that can respond to both electric and abrasive stimuli was successfully constructed by combining polymers and ILs.

3. Materials and Methods

The synthesis of the luminophore used in this study is shown in Scheme 1. NMR and IR spectra for each compound are provided in the Supporting Information.

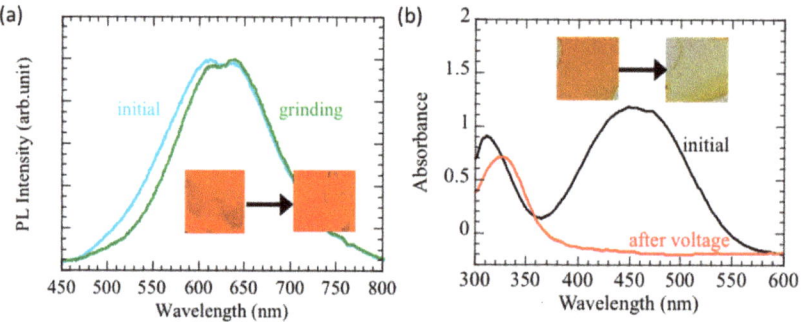

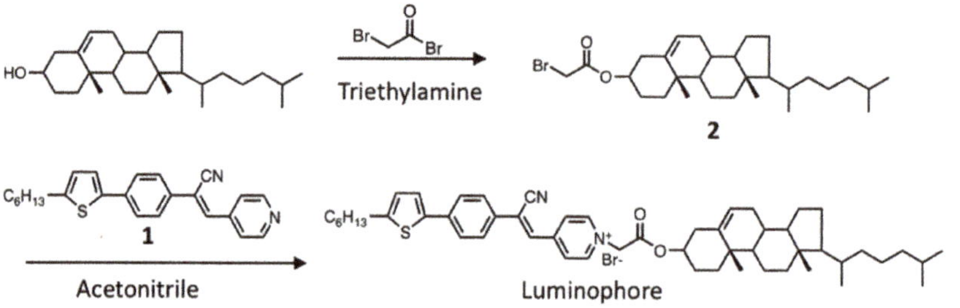

Scheme 1. Synthesis of the luminophore.

3.1. Synthesis

3.1.1. Synthesis of **1**

Compound **1** was synthesized according to a previously reported procedure [25].

^{1}H NMR (400 MHz, CDCl$_3$), δ (ppm): 0.91 (d, J = 7.5 Hz, 3H), 1.34–1.23 (m, 6H), 1.75–1.67 (m, 2H), 2.90–2.78 (m, 2H), 6.79 (d, J = 3.7 Hz, 1H), 7.26 (s, 2H), 7.47 (s, 1H), 7.66–7.62 (m, 2H), 7.68 (s, 1H), 7.75–7.71 (m, 2H), 8.77 (d, J = 15.3, 6.2 Hz, 2H).

FT-IR (KBr), ν (cm^{-1}): 3063, 2923, 2852, 2227, 1592, 1513, 1468, 1420, and 1002.

3.1.2. Synthesis of **2**

Cholesterol (3.82 g, 9.85 mmol) and triethylamine (8 mL) were dissolved in 20 mL THF in a 200 mL Erlenmeyer flask of Bromoacetyl bromide (2.1 g, 10.8 mmol) dissolved in 10 mL. THF was added dropwise to the reaction solution, and the reaction was stirred at room temperature for 3 h. After completion of the reaction, the by-products were removed by suction filtration, washed with THF, and the filtrate was collected. After reducing the filtrate under reduced pressure, the organic layer was washed with aqueous acetic acid and extracted with ethyl acetate. After dehydration with sodium sulfate, the filtrate was reduced under reduced pressure, purified by silica gel column chromatography, and recrystallized from THF/methanol to afford a yellow solid (1.79 g, M.p. 152 °C, 36% yield).

^{1}H NMR (400 MHz, CDCl$_3$): δ (ppm) 8.27 (d, J = 8.2 Hz, 2H), 8.02 (d, J = 8.2 Hz, 2H), 7.75 (d, J = 7.8 Hz, 2H), 7.70 (d, J = 8.2 Hz, 2H), 7.61 (d, J = 11.0 Hz, 3H), 7.47 (d, J = 7.8 Hz, 2H), 7.38 (d, J = 8.7 Hz, 2H), 7.20 (d, J = 7.8 Hz, 2H), 2.67 (q, J = 7.6 Hz, 2H), and 1.15–1.33 (3H).

IR (KBr, cm^{-1}): 2936, 1747, 1464, 1404, 1382, 1366, 1287, 1175, and 1026.

3.1.3. Synthesis of the Luminophore

0.16 g (0.4 mmol) of **1** and 0.31 g (0.06 mmol) of **2** were dissolved in 50 mL acetonitrile in a 200 mL round flask and stirred in an oil bath at 80 °C for 28 h. After the completion of the reaction, the filtrate was collected by suction filtration and washed with hexane to obtain a brown solid (0.08 g M.p. 150 °C, 21% yield).

^{1}H NMR (400 MHz, CDCl$_3$), δ (ppm): 0.71–1.66 (m, 5H), 1.13–0.85 (m, 41H), 1.29–1.38 (6H), 2.80–2.90 (2H), 3.81 (s, 2H), 6.09 (d, J = 5.5 Hz,1H), 6.79 (d, J = 3.7 Hz, 1H), 7.66 (d, J = 8.2 Hz, 2H), 7.85 (d, J = 8.7 Hz, 2H), 8.04 (s, 1H), 8.59 (d, J = 6.9 Hz, 2H), 9.25 (d, J = 6.9 Hz, 2H).

^{13}C NMR (101 MHz, CDCl$_3$): δ (ppm) 140.8, 121.8, 77.1, 76.8, 71.9, 56.8, 56.2, 50.2, 42.4, 39.8, 39.6, 37.3, 36.6, 36.3, 35.9, 32.0, 31.7, 31.7, 28.3, 28.1, 24.4, 23.9, 22.9, 22.7, 21.2, 18.8, and 11.9

FT-IR (KBr), ν (cm^{-1}): 3444, 2933, 1748, 1638, 1588, 1466, 1376, 1222, 1194, and 1056.

3.2. Equipment

The synthesized compounds were characterized using Fourier-transform nuclear magnetic resonance (FT-NMR; JEOL, Tokyo, Japan; JNM-ECZ 400 MHz) and infrared (FT-IR; JASCO, Tokyo, Japan; FT-IR 6000) spectroscopies. ^{1}H and ^{13}C NMR spectra were recorded for CDCl$_3$ at room temperature using tetramethylsilane (TMS; δ 0.00) as an internal standard. Polarized light microscopy (POM; Olympus, Tokyo, Japan; BH50) and differential scanning calorimetry (DSC; Hitachi High technologies, Tokyo, Japan; DSC7020) were used to evaluate the mesomorphic properties. PL spectra were collected using an emission spectrometer (Ocean Optics, Orlando, FL, US; USB2000+) equipped with a 365 nm UV LED as the excitation source (Ocean Optics, Orlando, US; LLS-LED). The solid-state quantum yields of the luminescent materials were measured using a fluorescence spectrometer (JASCO, Tokyo, Japan; FP-6600) equipped with an integrating sphere and excited at 360 nm. The PL lifetimes of the composite films were measured using a nanosecond spectrofluorometer (Horiba, Kyoto, Japan; FluoroCube) at an excitation wavelength of 370 nm. XRD measurements were performed using a diffractometer (Rigaku, Tokyo, Japan; SmartLab 3 kW) with a standard collimated beam setup. Microscopic images of xerogels were obtained using SEM (Keyence, Osaka, Japan; V8800). Thin layers for spectroscopy and XRD were prepared by spin-casting dioxane solutions onto glass substrates. Thin layers of the luminophores or polymer composites were ground in a pen, and the luminophore powder was ground in an agate mortar. DFT calculations were performed using Gaussian16 [26] at the B3LYP/6-31+G(d) level of theory. The absorption spectra were simulated using TD-DFT at the CAM-B3LYP/6-31+G(d) level of theory. The simulated absorption spectra were calculated for the first 10 excited states using the optimized ground-state geometry as the input geometry [27]. Cyclic voltammetry was performed using a voltammetry electrochemical analyzer (BAS, Tokyo, Japan; CV-50W) and a three-electrode cell with a carbon working electrode, a Pt counter electrode, and a Ag^+/Ag reference electrode. An NBu_4PF_6 solution (0.10 M) in DCM was used as the supporting electrolyte. All the potentials were calibrated using a ferrocene Fc^+/Fc couple as an internal reference. Polymer complexes were prepared using PMMA (182230-500G, Mw: 120,000 Da, T_g: 90 °C) purchased from Sigma-Aldrich without purification. The ILs and other solvents were purchased from TCI and used without further purification.

4. Conclusions

A luminescent dye with a quaternized pyridine terminus extended with cholesterol was synthesized, and its friability and electrical responsivity were evaluated. The dye exhibited a liquid crystalline phase at temperatures above 230 °C, and in the solid state, the maximum emission wavelength changed from orange emission at 637 nm to red emission at 654 nm. In complexation with EMI/TFSI, the film was gelatinized using chloroform/hexane as an auxiliary solvent; however, the deposited substrate was mechanically fragile and the MC response could not be confirmed. The dye was dissolved in a conductive gel composite of an IL and PMMA, and found that the dye exhibited a reversible sol–gel transition upon heating and cooling, forming a stable gel at room temperature. When the gel was deposited on a substrate and subjected to grinding, the emission wavelength changed from an orange-light emission wavelength of 611 nm to a red-light emission wavelength of 639 nm. Furthermore, when the gel was sandwiched between indium tin oxide electrodes and a reducing voltage of 3 V was applied, the coloration of the solution faded and, at the same time, the absorption at ~450 nm decreased, which is thought to be due to the desorption of cholesterol moieties by reduction.

Because the mechanisms of color change by electricity and luminescence color change by crushing are considered to be independent, a wider range of color changes can be expected when force and electricity are applied simultaneously. A material that can respond to electrical and mechanical stimuli can provide an intelligent skin that not only detects damage and overloads in flexible devices, but also conducts a certain amount of electricity and alerts the user to overvoltage points, such as those caused by short circuits. However, electro-responsiveness is irreversible with respect to thermoresponsiveness, and gels cannot be formed using ILs alone; these are issues that need to be addressed. In addition, as luminescent materials containing cyanostilbene are expected to change their liquid crystallinity [28] and emission color via photoreaction [29], the development of dyes that can respond to more stimuli is expected, and we plan to study these in the future.

Supplementary Materials: The following supporting information can be downloaded at: https://www.mdpi.com/article/10.3390/cryst13050786/s1, Figure S1: ^{1}H NMR Spectrum of the **1**; Figure S2: IR spectrum of **1**; Figure S3: ^{1}H NMR Spectrum of the **2**; Figure S4: IR spectrum of **2**; Figure S5: ^{1}H NMR Spectrum of the luminophore; Figure S6: IR spectrum of the luminophore; Figure S7: ^{13}C NMR Spectrum of the luminophore;.

Author Contributions: Conceptualization, N.K. and M.K.; methodology, M.K. and T.K.; software, M.K.; validation, S.Y., M.K., J.-i.N., and T.K.; investigation, S.Y. and S.S.; resources, N.K. and J.-i.N.; writing—original draft preparation, M.K.; writing—review and editing, N.K.; visualization, M.K.; supervision, N.K. All authors have read and agreed to the published version of the manuscript.

Funding: This study received no external funding.

Data Availability Statement: Data available in a publicly accessible repository.

Conflicts of Interest: The authors declare no conflict of interest.

References

1. Sagiri, S.S.; Behera, B.; Rafanan, R.R.; Bhattacharya, C.; Pal, K.; Banerjee, I.; Rousseau, D. Organogels as matrices for controlled drug delivery: A review on the current state. *Soft Mater.* **2014**, *12*, 47–72. [CrossRef]
2. Yu, X.; Chen, L.; Zhang, M.; Yi, T. Low-molecular-mass gels responding to ultrasound and mechanical stress: Towards self-healing materials. *Chem. Soc. Rev.* **2014**, *43*, 5346–5371. [CrossRef] [PubMed]
3. Chen, H.; Zhou, L.; Shi, X.; Hu, J.; Guo, J.; Albouy, P.A.; Li, M.H. AIE fluorescent gelators with thermo-, Mechano-, and Vapochromic properties. *Chem. Asian J.* **2019**, *14*, 781–788. [CrossRef]
4. Agarwal, D.S.; Prakash Singh, R.P.; Jha, P.N.; Sakhuja, R. Fabrication of deoxycholic acid tethered α-cyanostilbenes as smart low molecular weight gelators and AIEE probes for bio-imaging. *Steroids* **2020**, *160*, 108659. [CrossRef] [PubMed]
5. Cametti, M.; Džolić, Z. AIE-active supramolecular gel systems. In *Aggregation-Induced Emission (AIE); A Practical Guide*; Materials Today; Elsevier: Amsterdam, The Netherlands, 2022; pp. 117–164. [CrossRef]
6. Gao, A.; Wang, Q.; Wu, H.; Zhao, J.W.; Cao, X. Research progress on AIE cyanostilbene-based self-assembly gels: Design, regulation and applications. *Coord. Chem. Rev.* **2022**, *471*, 214753. [CrossRef]
7. Wang, Z.; Nie, J.; Qin, W.; Hu, Q.; Tang, B.Z. Gelation process visualized by aggregation-induced emission fluorogens. *Nat. Commun.* **2016**, *7*, 12033. [CrossRef]
8. Ku, K.; Hisano, K.; Yuasa, K.; Shigeyama, T.; Akamatsu, N.; Shishido, A.; Tsutsumi, O. Effect of Crosslinkers on Optical and Mechanical Behavior of Chiral Nematic Liquid Crystal Elastomers. *Molecules* **2021**, *26*, 6193. [CrossRef]
9. Stinson, V.P.; Shuchi, N.; McLamb, M.; Boreman, G.D.; Hofmann, T. Mechanical Control of the Optical Bandgap in One-Dimensional Photonic Crystals. *Micromachines* **2022**, *13*, 2248. [CrossRef]
10. van Esch, J.H.; Feringa, B.L. New functional materials based on self-assembling organogels: From serendipity towards design. *Angew. Chem. Int. Ed.* **2000**, *39*, 2263–2266. [CrossRef]
11. Seki, A.; Yoshio, M. Multi-color photoluminescence based on mechanically and thermally induced liquid-crystalline phase transitions of a hydrogen-bonded benzodithiophene derivative. *ChemPhysChem* **2020**, *21*, 328–334. [CrossRef] [PubMed]
12. Sagara, Y.; Kato, T. Brightly tricolored mechanochromic luminescence from a single-luminophore liquid crystal: Reversible writing and erasing of images. *Angew. Chem. Int. Ed. Engl.* **2011**, *50*, 9128–9132. [CrossRef] [PubMed]
13. Yamane, S.; Sagara, Y.; Mutai, T.; Araki, K.; Kato, T. Mechanochromic luminescent liquid crystals based on a bianthryl moiety. *J. Mater. Chem. C* **2013**, *1*, 2648–2656. [CrossRef]
14. Yagai, S.; Okamura, S.; Nakano, Y.; Yamauchi, M.; Kishikawa, T.; Kitamura, A.; Ueno, A.; Kuzuhara, D.; Yamada, H.; et al. Design amphiphilic dipolar p-systems for stimuli-responsive luminescent materials using metastable states. *Nat. Commun.* **2014**, *5*, 4013. [CrossRef]
15. Sha, J.; Lu, H.; Zhou, M.; Xia, G.; Fang, Y.; Zhang, G.; Qiu, L.; Yang, J.; Ding, Y. Highly polarized luminescence from an AIEE-active luminescent liquid crystalline film. *Org. Electron.* **2017**, *50*, 177–183. [CrossRef]
16. Kondo, M.; Yamoto, T.; Tada, M.; Kawatsuki, N. Mechanoresponsive behavior of rod-like liquid crystalline luminophores on an alignment layer. *Chem. Lett.* **2021**, *50*, 812–815. [CrossRef]
17. Panthai, S.; Fukuyama, R.; Hisano, K.; Tsutsumi, O. Stimuli-sensitive aggregation-induced emission of organogelators containing mesogenic Au(I) complexes. *Crystals* **2020**, *10*, 388. [CrossRef]
18. Kondo, M.; Morita, Y.; Kawatsuki, N. Directional blue-shifting Mechanofluorochromic luminescent behavior of liquid crystalline composite polymeric films. *Crystals* **2021**, *11*, 000950. [CrossRef]
19. Barbee, M.H.; Mondal, K.; Deng, J.Z.; Bharambe, V.; Neumann, T.V.; Adams, J.J.; Boechler, N.; Dickey, M.D.; Craig, S.L. Mechanochromic stretchable electronics. *ACS Appl. Mater. Interfaces* **2018**, *10*, 29918–29924. [CrossRef]
20. Sum, N.; Su, K.; Zhou, Z.; Wang, D.; Fery, A.; Lissel, F.; Zhao, X.; Chen, C. "Colorless-to-Black" Electrochromic and AIE-Active Polyamides: An Effective Strategy for the Highest-Contrast Electrofluorochromism. *Macromolecules* **2020**, *53*, 10117–10127. [CrossRef]
21. Moon, H.C.; Kim, C.-H.; Lodge, T.P.; Frisbie, D. Multicolored, Low-Power, Flexible Electrochromic Devices Based on Ion Gels. *ACS Appl. Mater. Interfaces* **2016**, *8*, 6252–6260. [CrossRef] [PubMed]
22. Moon, H.C.; Lodge, T.P.; Frisbie, C.D. Solution Processable, Electrochromic Ion Gels for Sub-1 V, Flexible Displays on Plastic. *Chem. Mater.* **2015**, *27*, 1420–1425. [CrossRef]
23. Zhang, Y.; Guo, M.; Li, G.; Chen, X.; Liu, Z.; Shao, J.; Huang, Y.; He, G. Ultrastable Viologen Ionic Liquids-Based Ionogels for Visible Strain Sensor Integrated with Electrochromism, Electrofluorochromism, and Strain Sensing. *CCS Chem. In press.* [CrossRef]
24. Ji, X.; Rosset, S.; Shea, H.R. Soft tunable diffractive optics with multifunctional transparent electrodes enabling integrated actuation. *Appl. Phys. Lett.* **2016**, *109*, 191901. [CrossRef]
25. Kondo, M.; Okumoto, K.; Miura, S.; Nakanishi, T.; Nishida, J.; Kawase, T.; Kawatsuki, M. Multicolor change in the photoluminescence induced by mechanical and chemical stimuli. *Chem. Lett.* **2017**, *46*, 1188–1190. [CrossRef]
26. Frish, M.J.; Trucks, G.W.; Schlegel, H.B.; Scuseria, G.E.; Robb, M.A.; Cheeseman, J.R.; Scalmani, G.; Barone, V.; Mennucci, B.; Petersson, G.A.; et al. *Gaussian16, Revision B.01*; Gaussian, Incorp.: Wallingford, CT, USA, 2016.
27. Shkoor, M.; Mehanna, H.; Shabana, A.; Farhat, T.; Bani-Yaseen, A.D. Experimental and DFT/TD-DFT computational investigations of the solvent effect on the spectral properties of nitro substituted pyridino [3,4-c] coumarins. *J. Mol. Liq.* **2020**, *313*, 113509. [CrossRef]

28. Liu, Z.; Liao, J.; He, L.; Gui, Q.; Yuan, Y.; Zhang, H. Preparation, photo-induced deformation behavior and application of hydrogen-bonded crosslinked liquid crystalline elastomers based on α-cyanostilbene. *Polym. Chem.* **2020**, *11*, 6047–6055. [CrossRef]
29. Ma, T.; Chen, S.; Du, X.; Mo, M.; Cheng, X. High-contrast fluorescence modulation driven by intramolecular photocyclization and protonation of bithienylpyridine functionalized α-cyanostilbene. *Dye. Pigment.* **2023**, *213*, 111176. [CrossRef]

Disclaimer/Publisher's Note: The statements, opinions and data contained in all publications are solely those of the individual author(s) and contributor(s) and not of MDPI and/or the editor(s). MDPI and/or the editor(s) disclaim responsibility for any injury to people or property resulting from any ideas, methods, instructions or products referred to in the content.

Article

Effects of Tetrafluorocyclohexa-1,3-Diene Ring Position on Photoluminescence and Liquid-Crystalline Properties of Tricyclic π-Conjugated Molecules

Haruka Ohsato, Shigeyuki Yamada *, Motohiro Yasui and Tsutomu Konno *

Faculty of Molecular Chemistry and Engineering, Kyoto Institute of Technology, Matsugasaki, Sakyo-ku, Kyoto 606-8585, Japan; m1673006@edu.kit.ac.jp (H.O.); myasui@kit.ac.jp (M.Y.)
* Correspondence: syamada@kit.ac.jp (S.Y.); konno@kit.ac.jp (T.K.)

Abstract: Tetrafluorocyclohexa-1,3-diene ring-containing tricyclic π-conjugated molecules are promising negative-dielectric-anisotropy guest species for vertical-alignment-type liquid-crystalline (LC) displays. Building on our previous work reporting the excellent photoluminescence (PL) properties of tricyclic π-conjugated molecules with central tetrafluorocyclohexa-1,3-diene rings, we herein synthesized four analogous molecules with terminal tetrafluorocyclohexa-1,3-diene rings from commercially available precursors and investigated the effects of substituent type and diene ring position on PL and LC properties using microscopic and spectroscopic methods. One of the prepared molecules exhibited a relatively planar molecular structure and formed herringbone-type aggregates via π/F and CH/π interactions instead of forming stacked aggregates via π/π stacking interactions, thus exhibiting relatively strong PL in solution and crystalline states. Moreover, the PL color of this compound depended on the electronic character of its terminal substituents along the long molecular axis. Of the four prepared species, two featured terminal ethyl groups and formed one or more LC phases. The PL properties of these phases indicated that the related phase transition induced changes in the aggregate structure, PL wavelength, and PL color. Our results expand the applicability of CF_2CF_2 moiety-containing tricyclic compounds as functional molecules for the fabrication of next-generation PL, LC, and PL-LC materials.

Keywords: fluorine; tetrafluorocyclohexa-1,3-diene; photoluminescence; liquid crystal; tricyclic molecule; aggregation

Citation: Ohsato, H.; Yamada, S.; Yasui, M.; Konno, T. Effects of Tetrafluorocyclohexa-1,3-Diene Ring Position on Photoluminescence and Liquid-Crystalline Properties of Tricyclic π-Conjugated Molecules. *Crystals* **2023**, *13*, 1208. https://doi.org/10.3390/cryst13081208

Academic Editor: Ingo Dierking

Received: 10 July 2023
Revised: 28 July 2023
Accepted: 29 July 2023
Published: 3 August 2023

Copyright: © 2023 by the authors. Licensee MDPI, Basel, Switzerland. This article is an open access article distributed under the terms and conditions of the Creative Commons Attribution (CC BY) license (https://creativecommons.org/licenses/by/4.0/).

1. Introduction

Fluorinated organic molecules have drawn much attention as the structural components of pharmaceuticals [1,2] and agrichemicals [3,4], as well as major constituents of liquid crystals [5–7] and optoelectronic materials [8,9]. This popularity is due to the unique properties of fluorine [10], namely, its highest electronegativity among all elements (4.0 on the Pauling scale), second smallest atomic radius (147 pm according to Bondi [11]), and high dissociation energy of C–F bonds (105.4 kcal·mol^{-1}). In view of these properties, the introduction of fluorine into molecular structures enhances latent functions or promotes the emergence of new ones and is, therefore, a powerful approach for the development of novel organic functional materials.

Our group has developed efficient and selective synthetic routes to various fluorinated organic molecules [12,13] including those exhibiting photoluminescence (PL) and liquid-crystalline (LC) properties [14]. The results obtained so far indicate that the introduction of fluorine atoms substantially increases PL intensity in the solid state and induces the emergence of mesophases between crystalline (Cry) and isotropic (Iso) phases.

Previously, we prepared a tricyclic molecule with a central tetrafluorocyclohexa-1,3-diene ring (**1a**) as a guest molecule with negative dielectric anisotropy to develop vertical alignment-type LC materials [15–18] and showed that **1a** exhibits blue PL in the crystalline and solution states. On this basis, we synthesized analogous tricyclic molecules with controlled electron density along the long molecular axis (**1b** and **1c**) and revealed that their PL behavior is greatly affected by the electron density distribution, which, in turn, is influenced by the electronic properties of terminal substituents (Figure 1) [19].

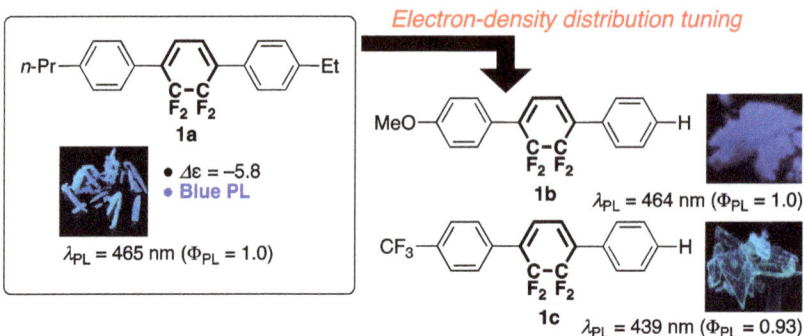

Figure 1. Previously synthesized tricyclic molecules **1a–c** with central tetrafluorocyclohexa-1,3-diene rings.

Building on the abovementioned results, we herein synthesized and characterized molecules **2a–d** to examine how PL and LC properties are affected by the position of the tetrafluorocyclohexa-1,3-diene ring in the tricyclic structure and the electronic properties of terminal substituents (Figure 2) [15,18].

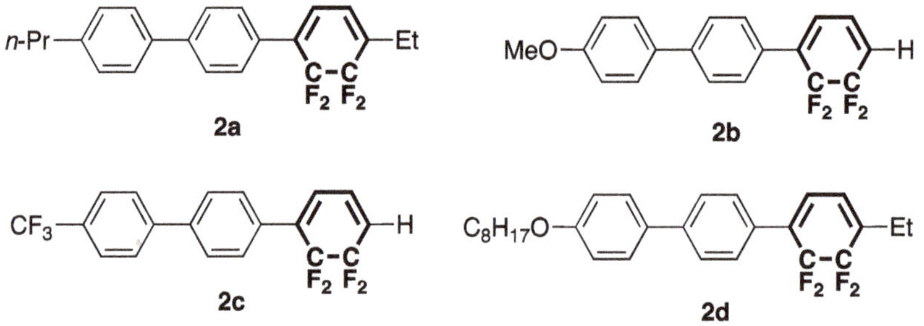

Figure 2. Structures of tricyclic molecules **2a–d** with terminal tetrafluorocyclohexa-1,3-diene rings.

2. Materials and Methods

2.1. General Characterization

Melting points (T_m) were measured on a Shimadzu DSC-60 Plus instrument using at least three heating/cooling cycles at a scan rate of 5.0 °C·min^{-1}. ^{1}H and ^{13}C nuclear magnetic resonance (NMR) spectra were recorded on a Bruker AVANCE III 400 spectrometer (^{1}H: 400.13 MHz, ^{13}C: 100.61 MHz) in chloroform-d (CDCl$_3$). Chemical shifts were reported on the basis of the residual proton or carbon signal of CHCl$_3$ (δ_H = 7.26 ppm, δ_C = 77 ppm) in parts per million (ppm). ^{19}F-NMR (376.46 MHz) spectra were recorded on the Bruker AVANCE III 400 spectrometer in CDCl$_3$ using trichlorofluoromethane (CFCl$_3$, δ_F = 0.00 ppm) as an internal standard. Infrared (IR) spectra were acquired using the KBr method on a JASCO FT/IR-4100 type A spectrometer. High-resolution mass spectra (HRMS) were recorded on a JEOL JMS-700MS spectrometer using fast atom bombardment (FAB+) methods. Column chromatography was performed using Wakogel® 60N (38–100 μm), and thin-layer chromatography was performed using the corresponding silica gel plates (silica gel 60F$_{254}$, Merck, Darmstadt, Germany).

2.2. Materials

The target molecules were synthesized according to a previously reported method from readily available precursors, namely, dimethyl 2,2,3,3-tetrafluorosuccinate (**2a** and **2d**; Scheme 1a) [18] and 4-bromo-3,3,4,4-tetrafluorobut-1-ene (**2b** and **2c**, Scheme 1b) [15].

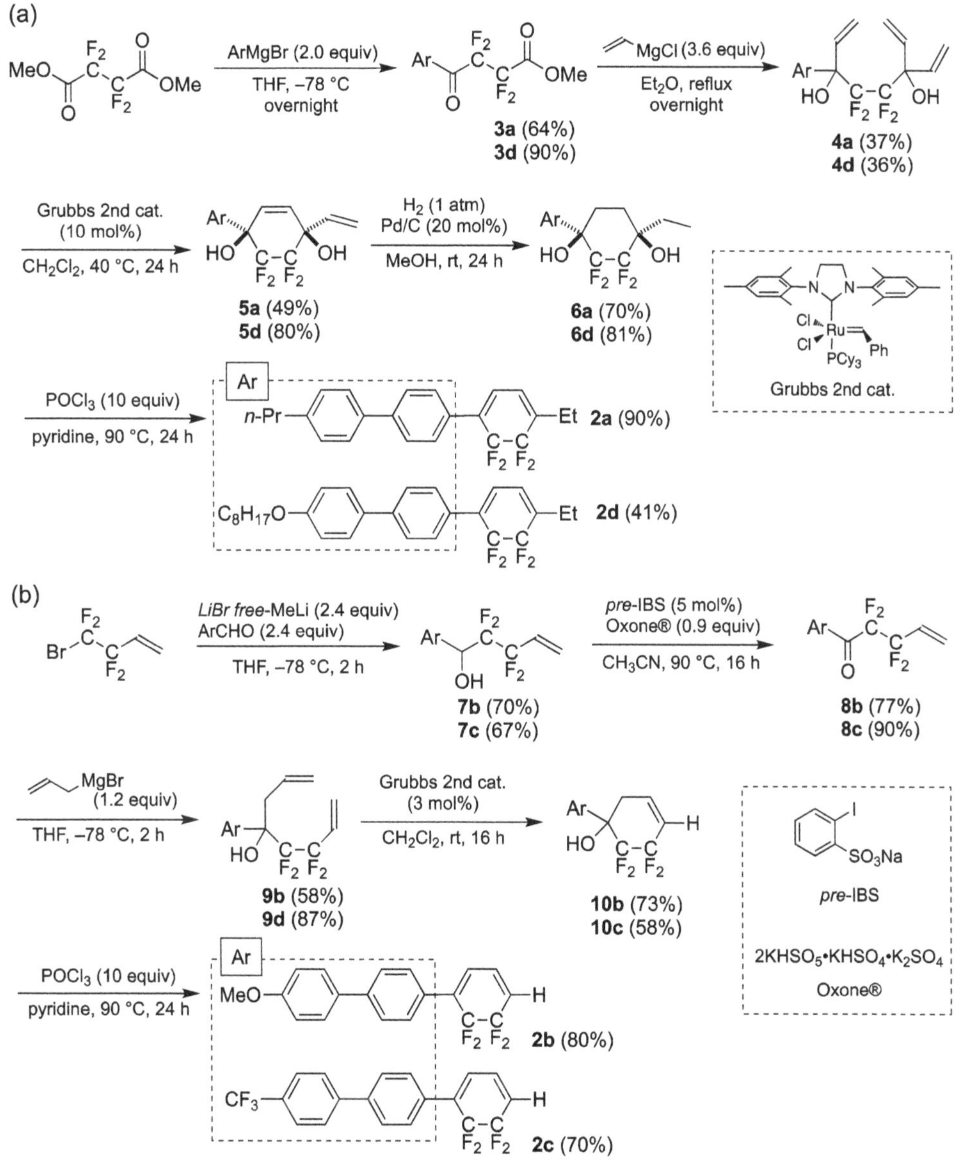

Scheme 1. Syntheses of (**a**) **2a** and **2d**, and (**b**) **2b** and **2c**.

Detailed synthetic procedures are provided in Schemes S1 and S2, and Figures S1–S42 shown in the Supplementary Materials. Characterization data are presented below (for **2a–d**) and in the Supplementary Materials (for other molecules).

2.2.1. 4-Ethyl-5,5,6,6-tetrafluoro-1-[4-(4-n-propylphenyl)phenyl]cyclohexa-1,3-diene (**2a**)

Yield: 90% (0.25 g, 0.67 mmol); yellow solid; T_m: 92 °C; ^{1}H-NMR (CDCl$_3$): δ 0.98 (t, J = 7.2 Hz, 3H), 1.20 (t, J = 7.6 Hz, 3H), 1.69 (sext, J = 7.6 Hz, 2H), 2.40 (q, J = 7.2 Hz, 2H), 2.64 (t, J = 7.2 Hz, 2H), 6.09 (d, J = 6.0 Hz, 1H), 6.39 (d, J = 6.0 Hz, 1H), 7.27 (d, J = 7.8 Hz, 2H), 7.53 (d, J = 8.4 Hz, 4H), 7.62 (d, J = 7.6 Hz, 2H); ^{13}C-NMR (CDCl$_3$): δ 11.4, 13.8, 21.6, 24.5, 37.7, 114.0 (tt, J = 251.87, 26.82 Hz), 114.1 (tt, J = 251.9, 26.8 Hz), 123.1 (t, J = 9.2 Hz), 125.8 (t, J = 8.79 Hz), 126.8, 126.9, 127.4, 129.0, 131.7, 133.6 (t, J = 22.0 Hz), 137.6, 137.8 (t, J = 21.9 Hz), 141.5, 142.3; ^{19}F-NMR (CDCl$_3$): δ −126.57 (d, J = 4.76 Hz, 2F), −122.23 (d, J = 4.86 Hz, 2F). The above characterization data were consistent with those reported previously [15,18].

2.2.2. 5,5,6,6-Tetrafluoro-1-(4-methoxyphenyl)phenylcyclohexa-1,3-diene (**2b**)

Yield: 80% (0.53 g, 1.6 mmol); white solid; T_m: 130 °C; ^{1}H-NMR (CDCl$_3$): δ 3.86 (s, 3H), 6.03–6.12 (m, 1H), 6.37–6.46 (m, 2H), 7.00 (d, J = 8.8 Hz, 2H), 7.51–7.63 (m, 6H); ^{13}C-NMR (CDCl$_3$): δ 55.3, 110.0–116.0 (m, 2C of CF$_2$CF$_2$), 114.3, 122.9 (t, J = 25.8 Hz), 124.7 (t, J = 8.0 Hz), 126.7, 127.7, 128.1, 129.9 (t, J = 11.8 Hz), 130.9, 132.6, 136.4 (t, J = 22.0 Hz), 141.6, 159.5; ^{19}F-NMR (CDCl$_3$): δ −121.29 (s, 2F), −121.67 (s, 2F); IR (KBr): ν 3026, 2968, 2844, 1649, 1604, 1576, 1530, 1445, 1399, 1312, 1289, 1202, 1183, 1021, 1011, 879, 787 cm^{-1}; HRMS (FAB) calculated for C$_{19}$H$_{14}$F$_4$O [M]$^+$: 334.0980, found: 334.0980. Crystal data for C$_{19}$H$_{14}$F$_4$O (M = 334.30 g/mol): orthorhombic, space group $P\,2_1\,2_1\,2_1$, a = 5.5586(7) Å, b = 9.2038(15) Å, c = 29.282(4) Å, α = 90°, β = 90°, γ = 90°, V = 1498.1(4) Å^3, Z = 4, T = 173 K, μ(MoKα) = 0.710 mm^{-1}, D_{calc} = 1.482 g/cm^3, 98,894 reflections measured (3.042° ≤ 2θ ≤ 27.480°), 7212 unique (R_{int} = 0.0476, R_{sigma} = 0.0950), which were used in all calculations. The final R_1 was 0.0631 ($I > 2\sigma(I)$) and wR_2 was 0.1269 (all data).

2.2.3. 5,5,6,6-Tetrafluoro-1-{4-(trifluoromethyl)phenyl}phenylcyclohexa-1,3-diene (**2c**)

Yield: 70% (0.15 g, 1.4 mmol); white solid; T_m: 138 °C; ^{1}H-NMR (CDCl$_3$): δ 6.06–6.18 (m, 1H), 6.40–6.54 (m, 2H), 7.60 (d, J = 8.8 Hz, 2H), 7.64 (d, J = 8.8 Hz, 2H), 7.72 (s, 4H); ^{13}C-NMR (CDCl$_3$): δ 112.7 (tt, J = 249.4, 27.2 Hz), 113.4 (tt, J = 253.0, 26.5 Hz), 124.2 (q, J = 272.1 Hz), 123.4 (t, J = 25.6 Hz), 125.5 (t, J = 8.0 Hz), 125.8 (q, J = 3.6 Hz), 127.3, 127.4, 127.9, 129.7 (q, J = 32.3 Hz), 129.8 (t, J = 11.8 Hz), 132.6, 136.1 (t, J = 22.7 Hz), 140.4, 143.6; ^{19}F-NMR (CDCl$_3$): δ −62.43 (s, 3F), −121.31 (s, 2F), −121.72 (s, 2F); IR (KBr): ν 3088, 2362, 1919, 1690, 1616, 1502, 1425, 1274, 1210, 968, 875, 794, 739, 729 cm^{-1}; HRMS (FAB) calculated for C$_{19}$H$_{11}$F$_7$ [M]$^+$: 372.0749, found: 372.0759.

2.2.4. 4-Ethyl-5,5,6,6-tetrafluoro-1-[4-{4-(n-octyloxy)phenyl}phenyl]cyclohexa-1,3-diene(**2d**)

Yield: 83% (0.32 g, 0.69 mmol); pale-yellow solid; T_m: 71 °C; ^{1}H-NMR (CDCl$_3$): δ 0.90 (t, J = 6.8 Hz, 3H), 1.19 (t, J = 7.2 Hz, 3H), 1.26–1.42 (m, 8H), 1.48 (quin, J = 8.0 Hz, 2H), 1.81 (quin, J = 7.2 Hz, 2H), 2.40 (q, J = 7.2 Hz, 2H), 4.00 (t, J = 6.8 Hz, 2H), 6.09 (d, J = 6.0 Hz, 1H), 6.38 (d, J = 6.0 Hz, 1H), 6.98 (d, J = 8.8 Hz, 2H), 7.48–7.62 (m, 6H); ^{13}C-NMR (CDCl$_3$):δ 11.5, 14.1, 21.7, 22.7, 26.1, 29.26, 29.31, 29.4, 31.8, 68.1, 110–125 (m, 2C of CF$_2$CF$_2$), 114.9, 123.1 (t, J = 8.8 Hz), 125.6 (t, J = 8.8 Hz), 126.7, 127.5, 128.0, 131.3, 132.6, 133.7 (t, J = 23.1 Hz), 137.8 (t, J = 21.2 Hz), 141.3, 159.1; ^{19}F-NMR (CDCl$_3$): δ −123.55 (s, 2F), −127.90 (s, 2F); IR (KBr): ν 3038, 2926, 2852, 1885, 1654, 1606, 1579, 1529, 1500, 1253, 1132, 907, 864 cm^{-1}; HRMS (FAB) calculated for C$_{28}$H$_{32}$F$_4$O [M]$^+$: 460.2389, found: 460.2382.

2.3. Single-Crystal X-ray Diffraction (XRD)

Single-crystal XRD patterns were recorded on an XtaLAB AFC11 diffractometer (Rigaku, Tokyo, Japan). The reflection data were integrated, scaled, and averaged using CrysAlisPro software (v. 1.171.39.43a; Rigaku Corporation, Akishima, Japan), and empirical absorption corrections were applied using the SCALE 3 ABSPACK scaling algorithm (CrysAlisPro). Structures were identified using a direct method (SHELXT-2018/2 [20]), refined using a full-matrix least-squares method (SHELXL-2018/3 [21]), and visualized using OLEX2 [22]. The crystallographic data were deposited in the Cambridge Crystallographic Data Center (CCDC) database (CCDC 2269760 for **2b**) and can be obtained free of charge from the CCDC, 12 Union Road, Cambridge CB2 1EZ, UK; Fax: +44-1223-336033; e-mail: deposit@ccdc.cam.ac.uk.

2.4. Photophysical Properties

JASCO V-750 absorption (JASCO, Tokyo, Japan) and FP-6600 fluorescence (JASCO, Tokyo, Japan) spectrometers were used to acquire solution-phase ultraviolet/visible (UV/vis) absorption and PL spectra. A Quantaurus-QY C11347-01 instrument (Hamamatsu Photonics, Hamamatsu, Japan) was used for PL quantum yield measurements, and a Quantaurus-Tau fluorescence lifetime spectrometer (C11367-34; Hamamatsu Photonics, Japan) was employed for PL lifetime determination.

2.5. LC Properties

Polarizing optical microscopy (POM) measurements were carried out using an Olympus BX53 microscope (Tokyo, Japan) equipped with cooling and heating stages (10,002 L, Linkam Scientific Instruments, Surrey, UK) to assess LC properties. Thermodynamic properties were assessed using differential scanning calorimetry (DSC; DSC-60 Plus, Shimadzu, Kyoto, Japan) at heating and cooling rates of 5.0 $°C·min^{-1}$ under N_2. Variable-temperature powder X-ray diffraction (VT-PXRD) analyses were carried out using an X-ray diffractometer (Rigaku, MiniFlex600, Tokyo, Japan) equipped with an X-ray tube (Cu K_α, λ = 1.54 Å) and semiconductor detector (D/teX Ultra2). The sample powder was mounted on a non-reflecting silicon plate set on a benchtop stage (Anton Paar, BTS-500). The temperature, heating/cooling rate, and X-ray exposure time were controlled.

2.6. Theoretical Calculations

All computations were performed using the Gaussian 16 program set [23] with density functional theory (DFT) at the level of the M06-2X hybrid functional [24] and the 6-31+G(d) (for all atoms) basis set with a conductor-like polarizable continuum model (CPCM) [25] for $CHCl_3$. Theoretical vertical transitions were calculated using the time-dependent DFT (TD-DFT) method at the same theoretical level using the same solvation model.

3. Results and Discussion
3.1. Synthesis

Compounds **2a** and **2d**, featuring an ethyl group attached to the longitudinal molecular terminal, were synthesized from the readily available dimethyl 2,2,3,3-tetrafluorosuccinate according to a reported procedure (Scheme 1a) [18]. The reaction of dimethyl tetrafluorosuccinate with 4-(4-*n*-propylphenyl)phenylmagnesium bromide in THF at -78 °C overnight followed by hydrolysis under acidic conditions afforded ketoester **3a** in 64% yield. Compound **3a** was treated with 3.6 equivalents of vinylmagnesium chloride in Et_2O, and the reaction mixture was stirred overnight at reflux to afford 4,4,5,5-tetrafluoroocta-1,7-diene (**4a**) in 37% yield. In the presence of a second-generation Grubbs catalyst, the ring-closing metathesis of **4a** in CH_2Cl_2 (40 °C, 24 h) furnished 1-aryl-4-ethyl-5,5,6,6-tetrafluorocyclohex-2-ene-1,4-diol (**5a**) in 49% yield. The 24 h exposure of **5a** in methanol to H_2 at room temperature resulted in catalytic hydrogenation and furnished 1-aryl-4-ethyl 2,2,3,3-tetrafluorocyclohexan-1,4-diol (**6a**) in 70% yield. Subsequent dehydration with phosphoryl chloride in pyridine at 90 °C for 24 h produced **2a** in 90% yield. The octyloxy chain-bearing structural analog **2d** was prepared by a similar procedure starting with the addition of 4-(4-octyloxyphenyl)phenylmagnesium bromide.

Compound **2b**, featuring an electron-donating methoxy group, and **2c**, featuring an electron-withdrawing trifluoromethyl (CF_3) group at the longitudinal molecular end, were synthesized according to a previously reported procedure (Scheme 1b) [15]. The Barbier-type nucleophilic addition of 1,1,2,2-tetrafluorobut-3-enyllithium (prepared in situ from 4-bromo-3,3,4-4-tetrafluorobut-1-ene and LiBr-free MeLi) to *p*-anisaldehyde in tetrahydrofuran (THF) at -78 °C for 2 h gave tetrafluorohomoallyl alcohol **7b** in 70% yield. The oxidation of **7b** with Oxone® in the presence of sodium 2-iodobenzenesulfonate (*pre*-IBS; 5 mol%) in acetonitrile at 90 °C for 16 h afforded 1-aryl-2,2,3,3-tetrafluoropent-4-en-1-one (**8b**) in 77% yield. Compound **8b** was treated with allylmagnesium bromide in THF at −78 °C for 2 h to produce 4-aryl-5,5,6,6-tetrafluoroocta-1,7-diene-4-ol (**9b**) in 58% yield. Compound **9b** underwent ring-closing metathesis upon treatment with a second-generation Grubbs catalyst (3 mol.%) to furnish 4-aryl-5,5,6,6-cyclohex-1-en-4-ol (**10b**) in 73% yield. The dehydration of **10b** with phosphoryl chloride in pyridine at 90 °C for 24 h produced the target methoxy-substituted species (**2b**) in 80% yield. The CF$_3$-substituted **2c** was synthesized using the same reaction sequence.

Compounds **2a–d** were purified by column chromatography (eluent: hexane/EtOAc = 3/1 for **2a** or 10/1 for **2b–d**) and recrystallization from a 1:1 (v/v) mixture of CH$_2$Cl$_2$ and hexane. The molecular structures of the target molecules were confirmed by NMR spectroscopy, IR spectroscopy, and HRMS, and the related purities were sufficient for photophysical and LC property analyses.

Among **2a–d**, only the methoxy-substituted **2b** furnished single crystals appropriate for X-ray crystallographic analysis upon recrystallization, whereas **2a**, **2c**, and **2d** did not furnish single crystals even after multiple recrystallizations. Figure 3 shows the crystal structure of **2b** obtained by X-ray structure analysis.

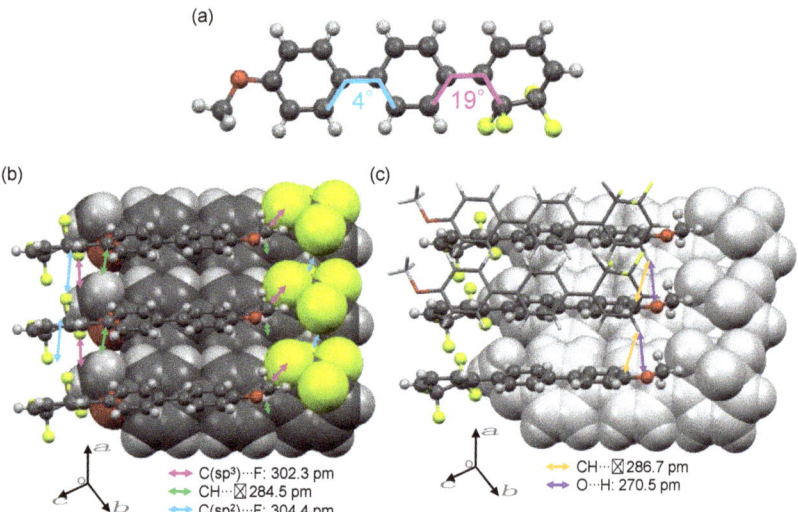

Figure 3. (**a**) Molecular structure and (**b**,**c**) packing of **2b** in the crystalline lattice. Display notation: space-filling model for rearmost molecules, ball-and-stick model for middle molecules, and wireframe model for frontmost molecules.

Compound **2b** crystallized in an orthorhombic system ($P\,2_1\,2_1\,2_1$ space group) and featured a unit cell with four molecules. The dihedral angle between the two aromatic rings of the biphenyl moiety was approximately 4°, and that between the tetrafluorocyclohexa-1,3-diene ring and the biphenyl moiety was approximately 19° (Figure 3a). In **1b**, which has a central tetrafluorocyclohexa-1,3-diene ring, the dihedral angle between the cyclohexa-1,3-diene ring and the adjacent aromatic ring was at least 31° [19]. On the basis of the molecular structures of **1b** and **2b**, we concluded that the change in the position of the tetrafluorocyclohexa-1,3-diene ring from central to terminal favored a more planar structure. The space-filling model representation in Figure 3b suggests that the π/F interactions [26,27] between the π-electrons and F atoms of tetrafluorocyclohexa-1,3-diene resulted in the formation of a stacked structure along the *a*-axis. The C(sp^2)···F interatomic distance corresponding to the π/F interaction (304.4 pm) was shorter than the sum of van der Waals radii (317 pm) of carbon (170 pm) and fluorine (147 pm) atoms [10]. The molecule represented by the space-filling model formed molecular packings featuring two pairs of CH/π interactions [28] with the molecule represented by the ball-and-stick model along the *b*-axis direction. The C(sp^2)···H interatomic distance corresponding to the CH/π interaction worked (284.5 pm) was also shorter than the sum of the van der Waals radii (290 pm) of carbon (170 pm) and hydrogen (120 pm). In addition to the short distance between the C(sp^2) and H atoms, the carbon atom of the methoxy group was in close contact with the fluorine atom at a distance (302.3 pm) shorter than the sum of carbon (170 pm) and fluorine (147 pm) van der Waals radii. The molecule represented by the ball-and-stick model also formed a stacked structure with the molecule represented by the wire-frame model along the *a*-axis via CH/π interactions (short contact: 286.7 pm) and O/H hydrogen bonds (short contact: 270.5 pm) (Figure 3c). Accordingly, herringbone-type packing structures were formed through multiple intermolecular interactions. However, unlike the packing structure of **1b** [19], which features a central cyclohexa-1,3-diene ring, the packing structure of **2b** did not feature intermolecular π/π stacking.

3.2. Photophysical Properties

Figure 4 shows the UV/vis absorption spectra, PL spectra, and PL color chromaticity diagrams (as defined by the Commission Internationale de l'Eclailage (CIE)) of **2a–d**, and Table 1 lists the related photophysical data.

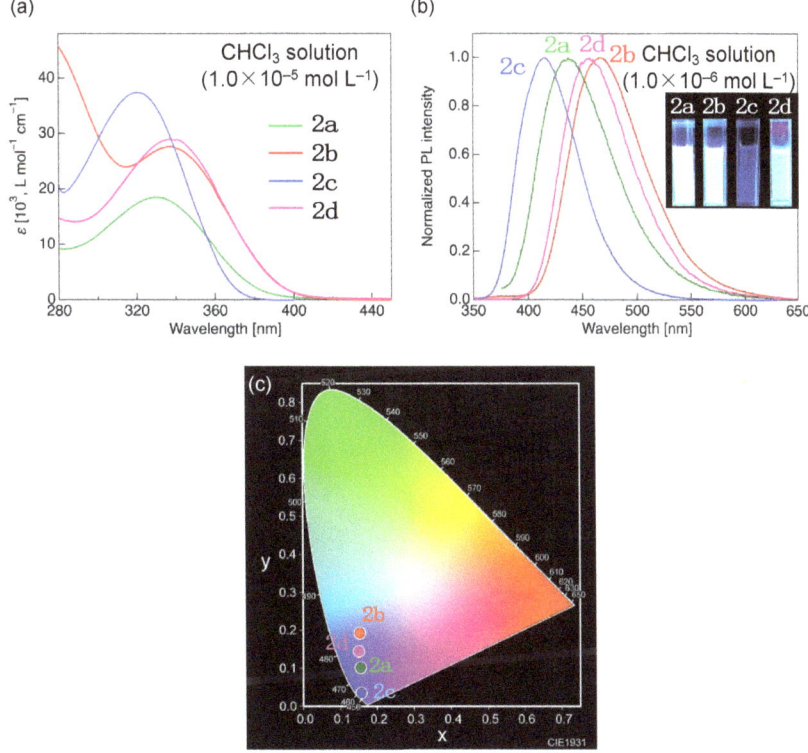

Figure 4. (**a**) Ultraviolet/visible absorption spectra (concentration: 1.0×10^{-5} mol·L^{-1}) and (**b**) photoluminescence (PL) spectra (concentration: 1.0×10^{-6} mol·L^{-1}) of **2a–d** measured in chloroform (CHCl$_3$). Inset: photographs of PL in CHCl$_3$ solution under UV irradiation (λ_{ex} = 365 nm). (**c**) Commission Internationale de l'Eclailage (CIE) chromaticity diagram for PL colors of **2a–d**.

Table 1. Photophysical data of **2a–d** in CHCl$_3$ solution.

Molecule	λ_{abs} [nm] [1] (ε [10^3, L·mol^{-1}·cm^{-1}])	λ_{PL} [nm] [2] (Φ_{PL}) [3]	τ [ns]	k_r [10^8, s^{-1}] [4]	k_{nr} [10^8, s^{-1}] [5]	CIE (x, y)
2a	330 (18.5)	437 (0.94)	2.08	4.52	0.28	(0.155, 0.102)
2b	337 (27.6)	463 (0.60)	1.84	3.29	2.15	(0.153, 0.193)
2c	320 (37.4)	416 (0.24)	0.73	3.30	10.44	(0.157, 0.036)
2d	337 (28.9)	463 (0.89)	2.11	4.21	0.53	(0.150, 0.145)

[1] Concentration: 1.0×10^{-5} mol·L^{-1}. [2] Concentration: 1.0×10^{-6} mol·L^{-1}. [3] Measured using an integrating sphere. [4] Radiative deactivation rate constant (k_r) = Φ_{PL}/τ. [5] Nonradiative deactivation rate constant (k_{nr}) = $(1 - \Phi_{PL})/\tau$.

Compound **2a**, possessing ethyl and *n*-propyl substituents at longitudinal molecular terminals, exhibited a single absorption band with a maximum absorption wavelength (λ_{abs}) of ~330 nm in CHCl$_3$. Compound **2b**, featuring a strongly electron-donating methoxy group, exhibited a red-shifted λ_{abs} of 337 nm, whereas **2c**, featuring a strongly electron-withdrawing CF$_3$ group, exhibited a blue-shifted λ_{abs} of 320 nm. In CHCl$_3$, the λ_{abs} of **2d** with ethyl and *n*-octyloxy groups as longitudinal terminal substituents was 337 nm, i.e., equal to that of **2b**.

The theoretical vertical transition was modeled using Gaussian software [23] with time-dependent density functional theory (TD-DFT). Figure 5 shows the distributions of the highest occupied molecular orbitals (HOMOs) and lowest unoccupied molecular orbitals (LUMOs) for **2a–d**, with the related theoretical data summarized in Table 2. The detailed orbital distributions are shown in Figures S44–47 in Supplementary Materials.

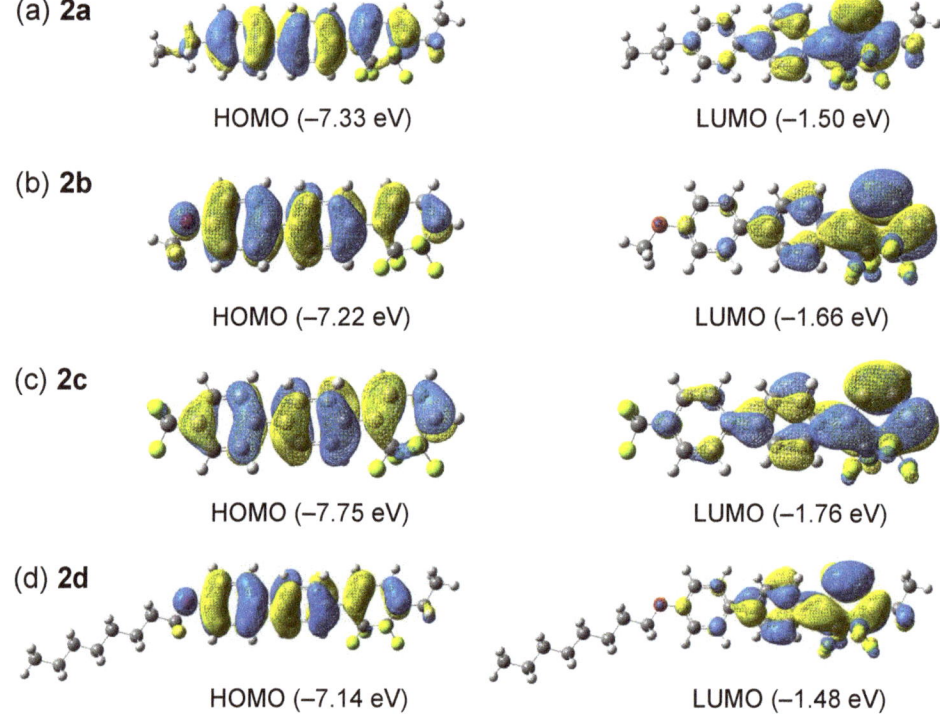

Figure 5. Highest occupied molecular orbital (HOMO, left) and lowest unoccupied molecular orbital (LUMO, right) distributions of (**a**) **2a**, (**b**) **2b**, (**c**) **2c**, and (**d**) **2d**.

Table 2. Theoretical data of **2a–d** obtained using Gaussian software with time-dependent density functional theory [1].

Molecule	HOMO Energy [eV]	LUMO Energy [eV]	Theoretical λ_{calcd} [nm]	Oscillator Strength (f)	Theoretical Transition (Probability)
2a	−7.33	−1.50	331	1.00	HOMO→LUMO (85%) HOMO−1→LUMO (12%)
2b	−7.22	−1.66	338	0.90	HOMO→LUMO (77%) HOMO−1→LUMO (20%)
2c	−7.75	−1.76	325	0.87	HOMO→LUMO (91%) HOMO−2→LUMO (6%)
2d	−7.14	−1.48	335	1.04	HOMO→LUMO (78%) HOMO−1→LUMO (19%)

[1] Calculated at the M06-2X/6-31+G(d) level of theory using a conductor-like polarizable continuum model for CHCl$_3$.

According to Figure 5, the HOMOs of **2a–d** were spread throughout the π-conjugated structure, whereas the LUMOs were localized on the tetrafluorocyclohexa-1,3-diene ring of the tricyclic π-conjugated framework. Substituents at longitudinal molecular ends affected the HOMO and LUMO energies, e.g., electron-donating substituents such as alkoxy groups increased the HOMO energy, whereas electron-withdrawing substituents had the opposite effect. The alkoxy group at the opposite end did not affect the energy of the LUMO, as this orbital was localized on the tetrafluorocyclohexa-1,3-diene ring, whereas the electron-donating ethyl group introduced into the tetrafluorocyclohexa-1,3-diene skeleton increased the LUMO energies of **2a** and **2d**. The absorption wavelengths (λ_{calcd}) of **2a–d** determined by TD-DFT calculations (331 nm for **2a**, 338 nm for **2b**, 325 nm for **2c**, and 335 nm for **2d**) were close to the measured λ_{abs} values listed in Table 1. The transitions from the ground to the first excited states were calculated to be of π–π* HOMO→LUMO and HOMO−1/HOMO−2→LUMO types.

When a solution of **2a** in CHCl$_3$ was excited by irradiation with UV light at λ_{abs} (330 nm), a single PL band with a maximum PL wavelength (λ_{PL}) of approximately 437 nm was observed (Figure 4b). Compared to that of **2a**, the PL band of **2b** with an electron-donating methoxy group (λ_{PL} = 463 nm in CHCl$_3$) was red-shifted by 26 nm, whereas the PL band of **2c** with an electron-withdrawing CF$_3$ group (λ_{PL} = 416 nm) was substantially blue-shifted. Similar to the methoxy-substituted **2b**, the *n*-octyloxy-substituted **2d** exhibited PL (λ_{PL} = 463 nm). In the case of **2c** with a large HOMO–LUMO overlap, radiative deactivation probably occurred from the locally excited state, whereas **2b** or **2d** with a locally existing LUMO luminesced through the radiative deactivation of the intramolecular charge transfer (ICT) excited state, which can be reasonably explained by the Lippert–Mataga plot [29,30] shown in Figure S49.

The high-energy PL of **2c** corresponded to dark-blue color represented by CIE chromaticity coordinates of (x, y) = (0.157, 0.036) (Figure 4c). In contrast, the low-energy PL of **2b** and **2d** emitted from ICT states corresponded to light-blue color with CIE coordinates of (x, y) = (0.153, 0.189). The quantum yields (Φ_{PL}) and PL lifetimes (τ) of **2a–d** were determined as 0.24–0.94 and ~2.11 ns, respectively. This value of τ indicates that the light emitted by **2a–d** was fluorescent. Among the four compounds, **2c** exhibited the lowest Φ_{PL} (0.24) and a very short τ (<1.0 ns). The radiative (k_r) and nonradiative (k_{nr}) deactivation rate constants of **2c** were calculated from Φ_{PL} and τ as 3.30×10^8 s^{-1} and 10.44×10^8 s^{-1}, respectively. Notably, k_r was not significantly different between **2a** and **d**, whereas the k_{nr} of **2c** was 5–37 times higher than those of other derivatives. These results suggested the occurrence of fluorescence reabsorption (self-absorption) in **2c**, which resulted in decreased Φ_{PL} and increased k_{nr}.

Most molecules exhibiting luminescence in solution generally experience luminescence quenching through intermolecular energy transfer at high concentrations or in the solid state. However, **2a–d** exhibited strong luminescence even in the crystalline state.

Figure 6 shows the PL spectra of crystalline **2a–d**, the related CIE chromaticity diagram, and photographs of crystals under 365 nm UV light. The corresponding photophysical data are summarized in Table 3.

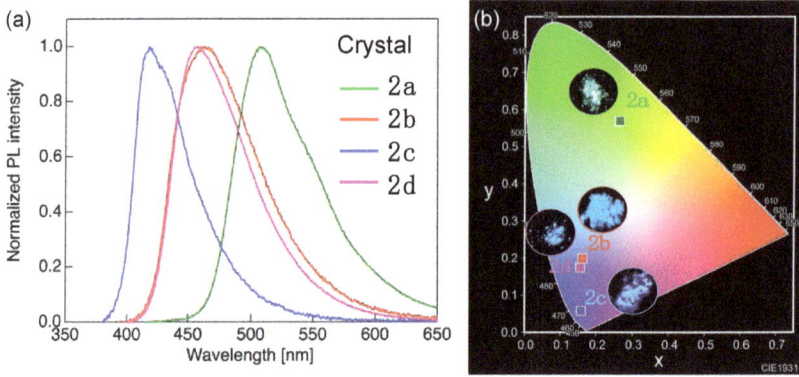

Figure 6. (a) PL spectra of crystalline **2a–d**. (b) CIE chromaticity diagram and photographs of **2a–d** crystals under 365 nm ultraviolet light.

Table 3. Photophysical data of crystalline **2a–d**.

Molecule	λ_{PL} [nm] (Φ_{PL}) [1]	τ_{ave} [ns]	τ_1 [ns]	τ_2 [ns]	k_r [10^8, s^{-1}] [2]	k_{nr} [10^8, s^{-1}] [3]	CIE (x, y)
2a	509 (0.99)	1.84	–	–	5.38	0.054	(0.265, 0.570)
2b	463 (0.63)	2.43	2.06	4.47	2.56	1.55	(0.162, 0.200)
2c	413 (0.31)	3.04	–	–	1.03	2.26	(0.155, 0.175)
2d	458 (0.93)	3.19	–	–	2.92	0.21	(0.182, 0.423)

[1] Measured using an integrating sphere. [2] Radiative deactivation rate constant (k_r) = Φ_{PL}/τ. [3] Nonradiative deactivation rate constant (k_{nr}) = $(1 - \Phi_{PL})/\tau$.

Crystalline **2a** with two alkyl groups at longitudinal molecular ends exhibited green PL with a single PL band at λ_{PL} around 509 nm, which was red-shifted relative to the value in CHCl$_3$ solution by 72 nm. However, the PL behavior of **2b–d** did not substantially change upon the transition from the CHCl$_3$ solution to the crystalline state. In the crystalline state, **2b** crystallized mainly via CH/π, π/F, and hydrogen bonds; π/π stacking between the intermolecular aromatic rings was not observed. The similarity between the λ_{PL} and Φ_{PL} values observed in the crystalline state and CHCl$_3$ solution was ascribed to the absence of π/π stacking interactions in crystalline **2b**, which suppressed the nonradiative deactivation induced by the formation of molecular aggregates. Given that crystalline **2c** and **2d** also exhibited PL behavior similar to that in the CHCl$_3$ solution state, we concluded that their conjugated structures were also not involved in intermolecular interactions, although their crystal structures have not yet been elucidated. On the other hand, we inferred that crystalline **2a**, which featured a PL wavelength and PL color different from those observed in the solution state, interacted with the π-conjugated site through the formation of molecular aggregates, unlike in the dilute solution, although the crystal structure of **2a** also remained veiled. PL lifetime measurements showed that the τ of crystalline **2a–d** was 1.84–3.19 ns and, therefore, also indicative of fluorescence. The PL decays of **2a**, **2c**, and **2d** were well modeled by a mono-exponential function, and the related PL originated from a single excited state. In contrast, the PL decay of **2b** was fitted by a biexponential function assuming a radiative deactivation pathway from any two excited states, although the related excited-state details remain unknown.

Compared with the previously reported **1b** and **1c** with a central tetrafluorocyclohexa-1,3-diene ring [19], **2b** and **2c** featured a shorter (by 20–25 nm) λ_{abs} in CHCl$_3$, which was ascribed to the significantly increased LUMO level of the latter molecules. However, λ_{PL} was found to be almost the same, except for **2c**, which had a CF_3 group at the molecular terminal. In the crystalline state, the λ_{PL} of **2c** was blue-shifted relative to that of **1c**, although almost identical λ_{PL} values were observed for **2b** and **1b**. The $CHCl_3$ solution-phase Φ_{PL} values of **2b** and **2c** exceeded those of **1b** and **1c**. In contrast, the opposite trend was observed in the crystalline state, i.e., the Φ_{PL} values of **2b** and **2c** were lower than those of **1b** and **1c**. In **2b** and **2c**, which greatly differ from **1b** and **1c** [19], the biphenyl moiety was planar and formed a herringbone structure because of CH/π interactions. We concluded that weak intermolecular interactions did not lead to molecular motion suppression, resulting in decreased Φ_{PL}. Accordingly, the positional change of the tetrafluorocyclohexa-1,3-diene ring in the tricyclic scaffold had a relatively large effect on the crystalline-state behavior, and the position of this ring affected intermolecular interactions and, hence, the extent of molecular motion inhibition and Φ_{PL}.

3.3. LC Properties

The previously reported **1a–c** exhibited transitions only between their Cry and Iso phases upon heating and cooling, i.e., no LC phases were observed [15]. To understand how the position of the tetrafluorocyclohexa-1,3-diene ring in tricyclic molecules affects their LC properties, we used POM and DSC to examine the LC behavior of **2a–d**, which showed PL in both dilute solution and crystalline states (Figures S53–S56 in Supplementary Materials). Compounds **2b** and **2c** exhibited only a Cry→Iso phase transition but did not form any mesophase upon heating and cooling. In contrast, for **2a** and **2d**, a fluid bright-field POM image was observed between the Cry and Iso phases, indicating the formation of an LC phase upon cooling (**2a**) or heating/cooling (**2d**). Figure 7 shows the DSC curves of **2a** and **2d** and the POM images of the corresponding mesophases. Table 4 lists the phase transition behaviors of **2a–d**, namely, their phase sequences, as well as phase transition temperatures and enthalpies in the second heating and cooling processes.

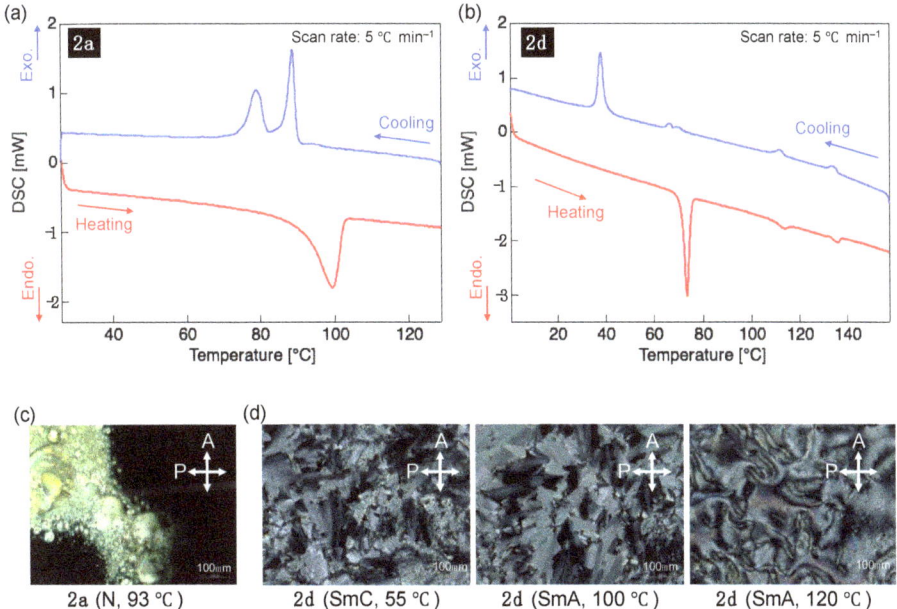

Figure 7. Differential scanning calorimetry (DSC) curves of (**a**) **2a** and (**b**) **2d** recorded during the second heating and cooling processes at a scan rate of 5 °C·min^{-1} under N_2. Polarizing optical microscopy textures in the mesophases of (**c**) **2a** and (**d**) **2d**.

Table 4. Phase transition data of **2a–d** during the second heating and cooling processes.

Molecule	Process	Phase Transition Temperatures [°C] and Enthalpies [kJ·mol^{-1}] [1]
2a	Heating	Cry 92 (14.0) Iso
	Cooling	Cry 81 (−4.9) N 90 (−6.2) Iso
2b	Heating	Cry 130 (16.4) Iso
	Cooling	Cry 93 (−12.1) Iso
2c	Heating	Cry 138 (11.8) Iso
	Cooling	Cry 98 (−8.2) Iso
2d	Heating	Cry 71 (12.6) SmA 110 (0.88) N 133 (0.72) Iso
	Cooling	Cry 40 (−8.1) SmC 68 (−1.1) SmA 114 (−1.0) N 136 (−0.85) Iso

[1] Determined by DSC (scan rate: 5 °C·min^{-1}, atmosphere: N$_2$). Abbreviations: Cry, crystal; Iso, isotropic; N, nematic; SmA, smectic A; SmC, smectic C phase.

In the case of the **2a** mesophase, POM revealed that a fluid four-brush Schlieren texture formed at 90 °C after the slow cooling from the dark-field-image Iso phase. Given that POM indicated the formation of a nematic (N) phase with only orientational order, the mesophase appearing during the cooling of **2a** was classified as the N phase. Further cooling from the N-phase state of **2a** resulted in fluidity loss at 81 °C and a phase transition to the hard Cry phase. In the case of the **2d** mesophase, the nonfluidic bright-field POM image corresponding to the Cry phase changed to a fluidic fan-shaped POM image at 71 °C upon heating. Further heating induced an optical texture change to a Schlieren-patterned N phase at 110 °C followed by a phase transition to the Iso phase in the dark-field POM image at 133 °C. Upon cooling, the N-phase Schlieren texture appeared at 136 °C, and a transition to a phase with a fan-shaped texture occurred at 114 °C. Upon further cooling, a broken fan-shaped texture was observed at 68 °C, followed by a phase transition to the nonfluidic Cry phase at 40 °C. The fan-shaped optical texture observed in the mesophase of **2d** is characteristic of the smectic (Sm) phase, which has an orientational and positional order. Notably, in the case of **2d**, the Sm phase appeared at a lower temperature than the N phase.

Further insights into the LC phases exhibited by **2a** and **2d** were provided by VT-PXRD measurements. The pattern of **2a** recorded after cooling from the Iso phase and holding at 70 °C featured no Cry phase peaks but contained a halo peak centered around $2\theta = 18°$ (Figure S57). This result strongly suggests that the mesophase appearing in **2a** is the N phase without positional order. PXRD measurements were also performed for **2d** at 124, 89, and 49 °C after cooling from the Iso phase. A halo peak centered around 18° was also observed in the pattern recorded at 124 °C, and the mesophase appearing at this temperature was determined to be the N phase (Figure S57). The PXRD pattern recorded at 89 °C featured a sharp peak at 3.75° and a weak peak at 7.45° (Figure 8a).

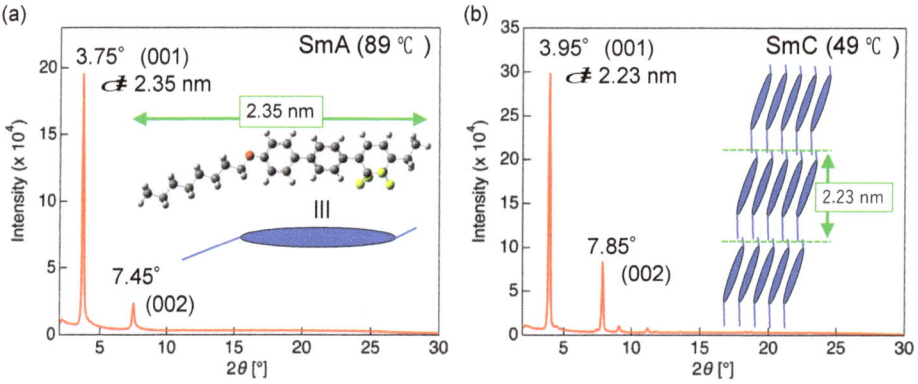

Figure 8. Powder X-ray diffraction patterns of **2d** recorded at (**a**) 89 and (**b**) 49 °C.

These diffraction peaks corresponded to the plane indices of (*hkl*) = (001) and (002). The peak at 3.75° in the low-angle region corresponded to a *d*-spacing of 2.35 nm, according to Bragg's equation, which was consistent with the longitudinal molecular length of **2d** (Figure 8a) This consistency of the interlayer distance with the molecular length agreed with the formation of a smectic A (SmA) phase with a layered periodic structure wherein the long molecular axis was oriented in the direction of the layer normal. In the pattern recorded at 49 °C, the peak of the (001) plane appeared at 3.95° and corresponded to a *d*-spacing of 2.23 nm, which was shorter than the molecular length along the long molecular axis (2.35 nm) (Figure 8b). This result indicated the presence of a smectic C (SmC) phase featuring a tilt angle with respect to the layer normal (Figure 8b).

3.4. PL Properties of **2d** in Various Molecular Aggregation States

Compound **2d**, which forms various mesophases, was selected to investigate PL behavior changes associated with the phase transition-induced alterations in molecular aggregate structure. PL behavior was examined using a fluorescence spectrometer equipped with a self-made temperature control unit. The samples were cooled from the Iso phase and held for 5 min at each temperature during cooling. Figure 9 shows the thus obtained PL spectra and CIE chromaticity diagrams, and Table 5 summarizes the related photophysical data.

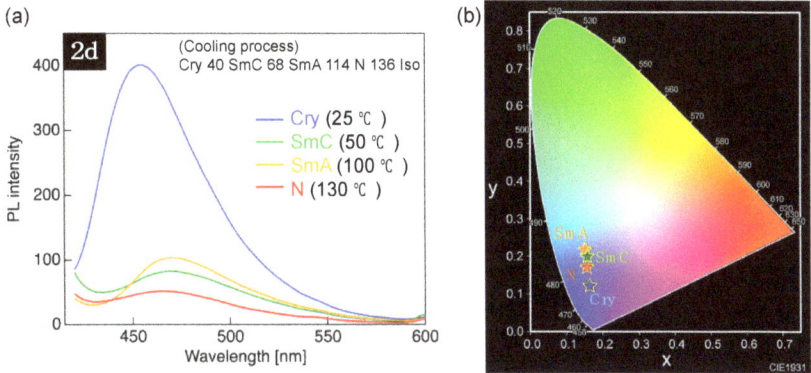

Figure 9. (a) PL spectra of **2d** recorded at different temperatures upon cooling. (b) CIE chromaticity diagram for PL color of **2d** at different temperatures.

Table 5. Photophysical data of **2d** in various phases.

Temp. [°C]/Phase	λ_{PL} [nm]	I/I_N [1]	CIE (*x*, *y*)
130/N	466	7.8	(0.156, 0.171)
100/SmA	470	1.6	(0.152, 0.217)
50/SmC	468	2.0	(0.158, 0.199)
25/Cry	454	1.0	(0.164, 0.121)

[1] PL intensity *I* of each phase with respect to the PL intensity (I_N) of the N phase.

In the case of **2d**, a PL band with $\lambda_{PL} \approx 466$ nm appeared in the N phase; however, the related PL intensity (I_N) decreased because of the accelerated nonradiative deactivation by micro-Brownian motion upon heating. The N→SmA phase transition observed upon cooling induced a 2.0-fold PL intensity increase (I/I_N = 2.0) along with a slight red shift in λ_{PL}. No significant change was observed in λ_{PL} or PL intensity upon the transition to the SmC phase, whereas the transition to the Cry phase induced a blue shift of λ_{PL} by 12 nm and a 7.8-fold increase in PL intensity relative to the N phase (I/I_N = 7.8). The CIE chromaticity diagram shown in Figure 9b demonstrates that the PL color of **2d** changed from dark blue to light blue owing to the phase transition-induced alteration of the molecular aggregate structure.

4. Conclusions

Tricyclic π-conjugated molecules with terminal tetrafluorocyclohexa-1,3-diene rings and different substituents introduced at the longitudinal molecular ends (**2a–d**) were synthesized in five steps from dimethyl 2,2,3,3-tetrafluorosuccinate or 4-bromo-3,3,4,4-tetrafluorobut-1-ene and evaluated in terms of their photophysical and LC behaviors. All four molecules exhibited PL in both dilute solutions and crystalline states. In dilute solutions, the PL wavelength varied in the range of 416–463 nm, which reflected the effect of substituent electron-donating/withdrawing nature on molecular orbital energy. Φ_{PL} was maximal (0.94) for **2a** and minimal (0.24) for **2c**, which had the shortest λ_{PL}. The low Φ_{PL} observed in the latter case was ascribed to self-absorption caused by the overlap of absorption and PL spectra. In the crystalline state, the PL behaviors of **2b–d** were similar to those in dilute solution, whereas **2a**, which had two alkyl groups at both ends, exhibited green PL with substantially red-shifted λ_{PL}. Regarding phase transition behavior, a mesophase was observed for **2a** and **2d** with an ethyl group at one molecular end. Only the N phase with an orientational order appeared in the case of **2a**, whereas the Sm phase with both orientational and positional orders, as well as the N phase, appeared in the case of **2d**. The N phase observed for **2d** exhibited weak blue PL during cooling. The PL intensity increased upon the N→SmA phase transition during cooling, did not substantially change upon the SmA→SmC phase transition, and strongly increased upon the SmC→Cry phase transition on further cooling. Concomitantly, the PL color changed from dark blue to light blue, i.e., temperature-responsive PL behavior was observed. The results described herein expand the applicability of CF_2CF_2-containing tricyclic molecules as next-generation PL, LC, and PL-LC materials.

Supplementary Materials: The following supporting information can be downloaded from https://www.mdpi.com/article/10.3390/cryst13081208/s1: Scheme S1. Synthetic procedure of **2a** and **2d** starting from commercially available dimethyl 2,2,3,3-tetrafluorosuccinate. Scheme S2. Synthetic procedure of **2b** and **2c** starting from commercially available 4-bromo-3,3,4,4-tetrafluorobut-1-ene. Figures S1–S42. ^{1}H-, ^{13}C-, and ^{19}F-NMR spectra; Figure S43. ORTEP-type crystal structure of **2b**; Figures S44–S47. HOMO-1/HOMO-2, HOMO, and LUMO distributions and differential density between HOMO and LUMO; Figure S48. UV/vis absorption and PL spectra of **2a–d** in CHCl$_3$ solution; Figure S49. PL spectra of **2a–c** in different solvents and related Lippert–Mataga plots; Figure S50. PL decay profiles of **2a–d** in CHCl$_3$ solution; Figure S51. Excitation and PL spectra of **2a–d** in crystalline states; Figure S52. PL decay profiles of **2a–d** in crystalline states; Figures S53–S56. DSC thermograms and POM images for **2a–d**; Figure S57. VT-PXRD patterns of **2a** and **2d** recorded at different temperatures; Table S1. Crystallographic data for **2b**; Tables S2–S5. Cartesian coordinates for **2a–d**; Tables S6–S9. Phase transition behaviors of **2a–d** observed by DSC.

Author Contributions: Conceptualization, H.O., S.Y., and T.K.; methodology, H.O., S.Y., and T.K.; validation, H.O., S.Y., and T.K.; formal analysis, H.O., S.Y., and T.K.; investigation, H.O., S.Y., and T.K.; resources, S.Y. and T.K.; data curation, H.O., S.Y., and T.K.; writing—original draft preparation, H.O., S.Y., and T.K.; writing—review and editing, H.O., S.Y., M.Y., and T.K.; Visualization, H.O., S.Y., and T.K.; supervision, T.K.; project administration, T.K.; funding acquisition, S.Y. and T.K. All authors have read and agreed to the published version of the manuscript.

Funding: This research received no external funding.

Data Availability Statement: Data supporting the presented findings are contained within the article and Supplementary Materials.

Acknowledgments: The authors express their sincere gratitude to Tosoh Finechem Corporation for providing 4-bromo-3,3,4,4-tetrafluorobut-1-ene and to Profs. Sakurai and Shimizu (Kyoto Institute of Technology) for help with VT-PXRD measurements. The authors acknowledge the use of equipment shared in the MEXT project to promote public utilization of advanced research infrastructure (program for supporting the introduction of the new sharing system), grant number JPMXS0421800222.

Conflicts of Interest: The authors declare no conflict of interest.

References

1. Inoue, M.; Sumii, Y.; Shibata, N. Contribution of organofluorine compounds to pharmaceuticals. *ACS Omega* **2020**, *5*, 10633–10640. [CrossRef] [PubMed]
2. Purser, S.; Moore, P.R.; Swallow, S.; Gouverneur, V. Fluorine in medicinal chemistry. *Chem. Soc. Rev.* **2008**, *37*, 320–330. [CrossRef] [PubMed]
3. Ogawa, Y.; Tokunaga, E.; Kobayashi, O.; Hirai, K.; Shibata, N. Current contributions of organofluorine compounds to the agrochemical industry. *iScience* **2020**, *23*, 101467. [CrossRef]
4. Jeschke, P. The unique role of fluorine in the design of active ingredients for modern crop protection. *ChemBioChem* **2004**, *5*, 570–589. [CrossRef]
5. Zhou, X.; Kang, S.W.; Kumar, S.; Li, Q. Self-assembly of discotic liquid crystal porphyrin into more controllable ordered nanostructure mediated by fluorophobic effect. *Liq. Cryst.* **2009**, *36*, 269–274. [CrossRef]
6. Wang, H.; Bisoyi, H.K.; Urbas, A.M.; Bunning, T.J.; Li, Q. Reversible circularly polarized reflection in a self-organized helical superstructure enabled by a visible-light-driven axially chiral molecular switch. *J. Am. Chem. Soc.* **2019**, *141*, 8078–8082. [CrossRef]
7. Wang, H.; Bisoyi, H.K.; Li, B.X.; McConney, M.E.; Bunning, T.J.; Li, Q. Visible-light-driven halogen bond donor based molecular switches: From reversible unwinding to handedness inversion in self-organized soft helical superstructures. *Angew. Chem. Int. Ed.* **2020**, *59*, 2684–2687. [CrossRef]
8. Kirsch, P. *Modern Fluoroorganic Chemistry: Synthesis, Reactivity, Applications*, 2nd ed.; Kirsh, P., Ed.; WILEY-VCH: Weinheim, Germany, 2013; pp. 247–298.
9. Hird, M. Fluorinated liquid crystals–properties and applications. *Chem. Soc. Rev.* **2007**, *36*, 2070–2095. [CrossRef]
10. O'Hagan, D. Understanding organofluorine chemistry. An introduction to the C–F bond. *Chem. Soc. Rev.* **2008**, *37*, 308–319. [CrossRef] [PubMed]
11. Bondi, A. van der Waals volumes and radii. *J. Phys. Chem.* **1964**, *68*, 441–451. [CrossRef]
12. Konno, T. Trifluoromethylated internal alkynes: Versatile building blocks for the preparation of various fluorine-containing molecules. *Synlett* **2014**, *25*, 1350–1370, and references cited therein. [CrossRef]
13. Ohsato, H.; Kawauchi, K.; Yamada, S.; Konno, T. Diverse synthetic transformations using 4-bromo-3,3,4,4-tetrafluorobut-1-ene and its applications in the preparation of CF_2CF_2-containing sugars, liquid crystals, and light-emitting materials. *Chem. Rec.* **2023**, *23*, e202300080, and references cited therein. [CrossRef]
14. Yamada, S.; Konno, T. Development of donor-π-acceptor-type fluorinated tolanes as compact condensed phase luminophores and applications in photoluminescent liquid-crystalline molecules. *Chem. Rec.* **2023**, *23*, e202300094, and references cited therein. [CrossRef] [PubMed]
15. Yamada, S.; Hashishita, S.; Asai, T.; Ishihara, T.; Konno, T. Design, synthesis and evaluation of new fluorinated liquid crystals bearing a CF_2CF_2 fragment with negative dielectric anisotropy. *Org. Biomol. Chem.* **2017**, *15*, 1495–1509. [CrossRef] [PubMed]
16. Yamada, S.; Hashishita, S.; Konishi, H.; Nishi, Y.; Kubota, T.; Asai, T.; Ishihara, T.; Konno, T. New entry for fluorinated carbocycles: Unprecedented 3,6-disubstituted 1,1,2,2-tetrafluorocyclohexane derivatives. *J. Fluorine Chem.* **2017**, *200*, 47–58. [CrossRef]
17. Yamada, S.; Tamamoto, K.; Kida, T.; Asai, T.; Ishihara, T.; Konno, T. Rational design and synthesis of a novel laterally-tetrafluorinated tricyclic mesogen with large negative dielectric anisotropy. *Org. Biomol. Chem.* **2017**, *15*, 9442–9454. [CrossRef] [PubMed]
18. Kumon, T.; Hashishita, S.; Kida, T.; Yamada, S.; Ishihara, T.; Konno, T. Gram-scale preparation of negative-type liquid crystals with a CF_2CF_2-carbocycle unit via an improved short-step synthetic protocol. *Beilstein J. Org. Chem.* **2018**, *14*, 148–154. [CrossRef]
19. Ohsato, H.; Morita, M.; Yamada, S.; Agou, T.; Fukumoto, H.; Konno, T. Aggregation-induced enhanced fluorescence by hydrogen bonding in π-conjugated tricarbocycles with a CF_2CF_2-containing cyclohexa-1,3-diene skeleton. *Mol. Syst. Des. Eng.* **2022**, *7*, 1129–1137. [CrossRef]
20. Sheldrick, G.M. SHELXT-integrated space-group and crystal-structure determination. *Acta Crystallogr. Sect. A Found. Adv.* **2015**, *71*, 3–8. [CrossRef]
21. Sheldrick, G.M. Crystal structure refinement with SHELXL. *Acta Crystallogr. Sect. C Struct. Chem.* **2015**, *71*, 3–8. [CrossRef] [PubMed]
22. Dolomanov, O.V.; Bourhis, L.J.; Gildea, R.J.; Howard, J.A.K.; Puschmann, H. OLEX2: A complete structure solution, refinement, and analysis program. *J. Appl. Crystallogr.* **2009**, *42*, 339–341. [CrossRef]
23. Frisch, M.J.; Trucks, G.W.; Schlegel, H.B.; Scuseria, G.E.; Robb, M.A.; Cheeseman, J.R.; Scalmani, G.; Barone, V.; Petersson, G.A.; Nakatsuji, H.; et al. *Gaussian 16, Revision B.01*; Gaussian, Inc.: Wallingford, CT, USA, 2016.
24. Hohenstein, E.G.; Chill, S.T.; Sherrill, C.D. Assessment of the performance of the M05-2X and M06-2X exchange-correlation functionals for noncovalent interactions in biomolecules. *J. Chem. Theory Comput.* **2008**, *4*, 1996–2000. [CrossRef] [PubMed]
25. Li, H.; Jensen, J.H. Improving the efficiency and convergence of geometry optimization with the polarizable continuum model: New energy gradients and molecular surface tesselation. *J. Comput. Chem.* **2004**, *25*, 1449–1462. [CrossRef] [PubMed]
26. Rybalova, T.V.; Bagryanskaya, I.Y. C–F···π, F···H, and F···F intermolecular interactions and F-aggregations: Role in crystal engineering of fluoroorganic compounds. *J. Struct. Chem.* **2009**, *50*, 741–753. [CrossRef]
27. Kawahara, S.; Tsuzuki, S.; Uchimaru, T. Theoretical study of the C–F/π interaction: Attractive interaction between fluorinated alkane and an electron-deficient π-system. *J. Phys. Chem. A* **2004**, *108*, 6744–6749. [CrossRef]

28. Tsuzuki, S.; Fujii, A. Nature and physical origin of CH/π interaction: Significant difference from conventional hydrogen bonds. *Phys. Chem. Chem. Phys.* **2008**, *10*, 2584–2594. [CrossRef] [PubMed]
29. Mataga, N.; Kaifu, Y.; Koizumi, M. The solvent effect on the fluorescence spectrum changes of solute-solvent interactions during the lifetime of the excited solute molecule. *Bull. Chem. Soc. Jpn.* **1955**, *28*, 690–691. [CrossRef]
30. Mataga, N.; Kaifu, Y.; Koizumi, M. Solvent effects on fluorescence spectra and dipole moments of excited molecules. *Bull. Chem. Soc. Jpn.* **1956**, *29*, 465–470. [CrossRef]

Disclaimer/Publisher's Note: The statements, opinions and data contained in all publications are solely those of the individual author(s) and contributor(s) and not of MDPI and/or the editor(s). MDPI and/or the editor(s) disclaim responsibility for any injury to people or property resulting from any ideas, methods, instructions or products referred to in the content.

Article

Side-Chain Labeling Strategy for Forming Self-Sorted Columnar Liquid Crystals from Binary Discotic Systems

Tsuneaki Sakurai [1,*], Kenichi Kato [2] and Masaki Shimizu [1]

[1] Faculty of Molecular Chemistry and Engineering, Kyoto Institute of Technology, Hashikami-cho, Matsugasaki, Sakyo-ku, Kyoto 606-8585, Japan
[2] RIKEN SPring-8 Center, 1-1-1 Kouto, Sayo-cho, Sayo-gun 679-5148, Japan
* Correspondence: sakurai@kit.ac.jp

Abstract: The spontaneous formation of self-sorted columnar structures of electron-donating and accepting π-conjugated molecules is attractive for photoconducting and photovoltaic properties. However, the simple mixing of donor–acceptor discotic molecules usually results in the formation of mixed-stacked or alternating-stacked columns. As a new strategy for overcoming this problem, here, we report the "side-chain labeling" approach using binary discotic systems and realize the preferential formation of such self-sorted columnar structures in a thermodynamically stable phase. The demonstrated key strategy involves the use of hydrophobic and hydrophilic side chains. The prepared blend is composed of liquid crystalline phthalocyanine with branched alkyl chains (H_2Pc) and perylenediimide (PDI) carrying alkyl chains at one side and triethyleneglycol (TEG) chains at the other side ($PDI_{C12/TEG}$). To avoid the thermodynamically unfavorable contact among hydrophobic and hydrophilic chains, $PDI_{C12/TEG}$ self-assembles to stack up on top of each other and H_2Pc as well, forming a homo-stacked pair of columns (self-sort). Importantly, H_2Pc and $PDI_{C12/TEG}$ in the blend are macroscopically miscible and uniform, and mesoscopically segregated. The columnar liquid crystalline microdomains of H_2Pc and $PDI_{C12/TEG}$ are homeotropically aligned in a glass sandwiched cell. The "labeling" strategy demonstrated here is potentially applicable to any binary discotic system and enables the preferential formation of self-sorted columnar structures.

Keywords: self-sort; segregated columns; binary mixture; amphiphilicity; homeotropic alignment

Citation: Sakurai, T.; Kato, K.; Shimizu, M. Side-Chain Labeling Strategy for Forming Self-Sorted Columnar Liquid Crystals from Binary Discotic Systems. *Crystals* **2023**, *13*, 1473. https://doi.org/10.3390/cryst13101473

Academic Editor: Benoit Heinrich

Received: 27 August 2023
Revised: 6 October 2023
Accepted: 7 October 2023
Published: 10 October 2023

Copyright: © 2023 by the authors. Licensee MDPI, Basel, Switzerland. This article is an open access article distributed under the terms and conditions of the Creative Commons Attribution (CC BY) license (https:// creativecommons.org/licenses/by/ 4.0/).

1. Introduction

The control of nanostructures and miscibility of binary blends is important for tuning the physical properties of the blended organic materials. Historically, polymer blend has been a famous notion, where the miscibility and compatibility of blended polymers have been well discussed, especially in view of their effect on thermal and mechanical properties [1–4]. The bulk heterojunction of conjugated polymers and fullerene derivatives is another famous concept utilized for blend films in organic photovoltaic cells [5–7]. More recently, blends of electron donors and acceptors based on conjugated polymers or small molecules have been used for active layers in organic electronic devices, including photovoltaic cells [8–10], electrochemical transistors [11], ambipolar transistors [12], and so on. In these blends, not only large interfaces of donor and acceptor molecules (or macromolecules) but also hole/electron-transporting bicontinuous interpenetrating networks are essential for the device operation. The optimization of such nanostructures in the blends is usually performed by a try-and-error approach using spin-coating methods. Meanwhile, hydrogen-bond-assisted organogelator systems have been demonstrated as a more elaborated molecular design [13–16]. In these systems, self-sorted fibrous one-dimensional assemblies were developed by the simple mixing of electron donor and acceptor molecules. The different distances of two hydrogen bonding sites between the donor and acceptor molecules are critical for the formation of self-sorting fibers. Although methodologies of bulk heterojunctions as well as binary organogelator systems have been established, they have a critical drawback: the obtained nanostructures are kinetically controlled but not thermodynamically stable in the long term. Thus, the construction of thermodynamic bicontinuous structures of electron donor and acceptor materials has been awaited.

Liquid crystal (LC) phases are usually thermodynamically stable and appropriate as a platform for constructing the arrays of π-conjugated systems through self-assembly, resulting in the functional soft materials [17–23]. However, there has been no example of binary LC blends with bicontinuous structures. In columnar LC phases of discotic π-systems, columnarly stacked π-conjugated molecules enable one-dimensional charge transport pathways. If electron-donating and accepting π-conjugated molecules form a self-sorted columnar structure, we can realize intracolumnar hole/electron transport pathways as well as intercolumnar p/n heterojunctions with large interfaces, which equips a long-term structural stability. However, in relevant previous studies, mixed-stacking columnar structures were reported from LC phthalocyanine (p-type: electron donor) and perylene diimide (PDI) (n-type: electron acceptor) molecules [24–26]. This is quite reasonable because they are entropically favored (Figure 1a). Special molecular designs are required to accomplish thermodynamically stable self-sorted columnar structures by the self-assembly of LC electron donor and acceptor molecules. Here, we report a "side-chain labeling" strategy to realize self-sorted columnar structures from LC mixtures composed of free-base phthalocyanine (**H_2Pc**) and PDI derivatives **$PDI_{C12/C12}$**, **$PDI_{C12/TEG}$**, and **$PDI_{TEG/TEG}$** (Figure 2). The dissymmetric introduction of both hydrophobic and hydrophilic side chains in **$PDI_{C12/TEG}$** gives large enthalpic gain to the homo-stacked PDI columns, and thus the self-sorted structure is stabilized when the **$PDI_{C12/TEG}$** is mixed with **H_2Pc** carrying hydrophobic chains (Figure 1b). Furthermore, the resulting self-sorted LC columns of the **$PDI_{C12/TEG}$** and **H_2Pc** molecules align homeotropically in a sandwiched glass cell, which is desirable for efficient charge transport in photovoltaic applications.

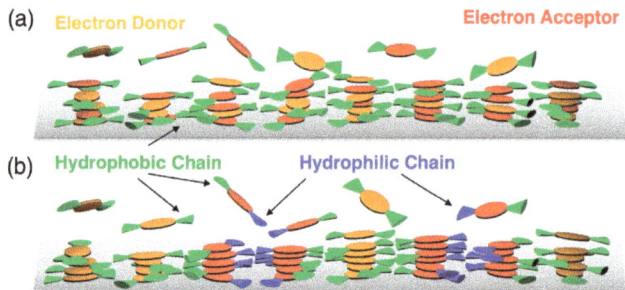

Figure 1. Schematic illustrations of side-chain-directed molecular assembly of electron donors (orange) and acceptor (red) π-systems (**a**) substituted with hydrophobic side chains (green) alone and (**b**) site-specifically substituted with hydrophobic (green) and hydrophilic (blue) side chains.

Entropically favored mixed-stacking structures can be formed because both electron donor and acceptor molecules are decorated with alkyl chains and they are molecularly miscible. We focused on strong enthalpic interactions of immiscible hydrophobic and hydrophilic chains. If donor (acceptor) molecules are substituted with alkyl chains and acceptor (donor) molecules with hydrophilic oxyethylene chains, they are not miscible but macroscopically segregated [27]. How do we access the thermodynamic self-sorted nanostructure? The clue is hidden in the amphiphilic molecular design used in previous works, including ours [28–32]. **$PDI_{C12/TEG}$** (Figure 2) is a Janus-type amphiphilic compound forming a columnar LC phase at room temperature. **$PDI_{C12/TEG}$** molecules pack into the rectangular columnar phase with $p2mg$ symmetry to minimize the unfavorable contact among immiscible hydrophobic and hydrophilic side chains incorporated in a single PDI core. When **H_2Pc**, a compound carrying hydrophobic chains, is blended with **$PDI_{C12/TEG}$**, **H_2Pc** molecules may not intercalate into a column of **$PDI_{C12/TEG}$** to avoid the enthalpic penalty of increasing contacts between hydrophobic and hydrophilic segments. Nevertheless, they are macroscopically miscible with one another due to their hydrophobic chains. We expected that **H₂Pc** molecules would form a homo-stacked columnar assembly and laterally contact with the hydrophobic chains of **PDI$_{C12/TEG}$**, resulting in the self-sorted columnar assembly (Figure 1b). The preferential formation of a self-sorted nanostructure will be discussed in detail in the section of Results and Discussion.

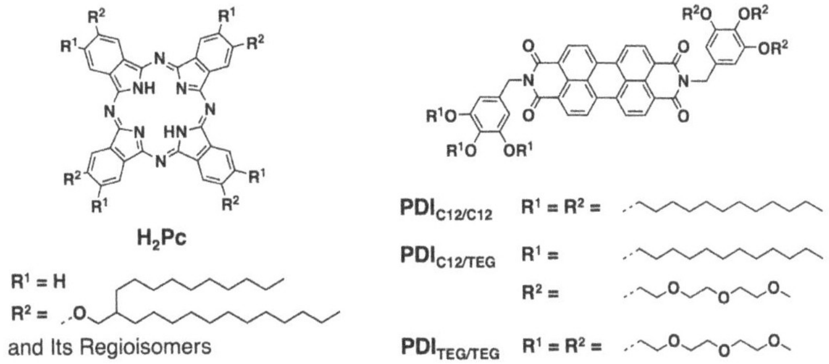

Figure 2. Chemical structures of liquid crystalline phthalocyanine **H₂Pc** and perylenediimides **PDI$_{C12/C12}$**, **PDI$_{C12/TEG}$**, and **PDI$_{TEG/TEG}$**.

2. Materials and Methods

2.1. Synthesis and Characterization of H₂Pc and PDIs

H₂Pc, **PDI$_{C12/C12}$**, **PDI$_{C12/TEG}$**, and **PDI$_{TEG/TEG}$** were synthesized according to the previous reports [31,33], and characterized by ^{1}H NMR spectroscopy in CDCl$_3$ on a Varian model Mercury 400 spectrometer, operating at 400 MHz, where chemical shifts were determined with respect to tetramethylsilane as an internal reference. MALDI-TOF mass spectrometry was performed on an Autoflex III spectrometer from Bruker, Japan, using dithranol as a matrix. In addition, 1:1 molar ratio mixtures of **H₂Pc** and PDI derivatives were prepared from their CH$_2$Cl$_2$ solutions in a glass vial. The solvent was evaporated in each solution, allowing for as-prepared waxy LC mixture.

2.2. Characterization of LC Mesophases

The optical textures were recorded by a BX53-P polarizing optical microscope (POM) from Olympus, Japan, equipped with an EOS kiss X7i digital camera from Canon, Japan. The sample was loaded, by use of a capillary action, into an LC cell without any surface treatment. The LC cell was prepared as follows. The glasses with a size of 16 × 22 × 0.5 mm were purchased from Matsunami Glass Ind., Ltd. (Haemacytomer Cover Glasses), Osaka, Japan. Silica beads with 5 µm diameter were dispersed in a drop of fast curing optical adhesive (NOA81) purchased from THORLABS, and the beads-dispersed adhesive was spotted at four places in a rectangle on one glass. Another glass was placed onto the adhesive-spotted glass with a few millimeter offset along with the long axis. The sandwiched glass cell was irradiated with 365 nm light from a SLUV-4 handy UV lamp purchased from AS ONE, Japan, to complete the curing of the adhesive.

The temperature of the sample was controlled by a HS82 hot-stage from Mettler Toledo, Japan. Differential scanning calorimetry (DSC) measurements were performed on a DSC 822e differential scanning calorimeter from Mettler Toledo, Japan. Cooling and heating profiles were recorded and analyzed with the STARe system. Samples were put into an aluminum pan and allowed to be measured under N$_2$ gas flow.

X-ray diffraction measurements were carried out using a synchrotron radiation X-ray beam with a wavelength of 0.108 nm on BL44B2 at the Super Photon Ring (SPring-8, Hyogo, Japan) [34]. A large Debye–Scherrer camera was used in conjunction with an imaging plate as a detector, and all diffraction patterns were recorded with a 0.01° step in 2θ. The samples were loaded by capillary action at the isotropic liquid melts into a 0.5 mm thick soda glass capillary purchased from WJM-Glas/Muller GmbH. During the measurements, samples were continuously rotated along the capillary axis to obtain a homogeneous diffraction pattern. The exposure time to the X-ray beam was 1.5 min each.

2.3. Evaluation of Intracolumnar Molecular Order

Electronic absorption spectra were recorded on a V-730 UV/VIS/NIR spectrophotometer from JASCO, Japan, where the scan rate, response, and band width were set at 1000 nm min^{-1}, 0.06 s, and 1.0 nm. The CHCl$_3$ solution samples were prepared at 2.0×10^{-5} M and measured in a quartz cell equipped with a screw cap. The optical path length of the cell is 1.0 cm. Spin-coated films were prepared from CHCl$_3$ solutions of the single compound or 1:1 molar ratio **H$_2$Pc**/PDI mixtures onto a quartz substrate with a size of $9 \times 40 \times 1$ mm. The spin-coating was performed at 1500 rpm for 30 s using a Mikasa model MS B-100 spin coater.

3. Results and Discussion

3.1. Homeotropic Alignment Capability of H$_2$Pc and PDIs

The phase transition behaviors of **H$_2$Pc**, **PDI$_{C12/C12}$**, **PDI$_{C12/TEG}$**, and **PDI$_{TEG/TEG}$** were characterized by DSC (Figure S1). They all showed LC mesophases and their clearing points are 180, 223, 189, and 165 °C on cooling, respectively, which is almost identical with the previous reports [31,33]. A spontaneous homeotropic alignment of discotic columnar LCs was often reported for hexagonal columnar mesophases [35–38]. The homeotropic alignment capability of **H$_2$Pc** discotic columns was already reported in a previous study [39]. The capability of spontaneous homeotropic alignment for the PDI derivatives was monitored by means of POM using samples.

After being loaded into the glass cell with a capillary action at the isotropic liquid phase (Iso), the sample was slowly cooled at 1.0 K/min. Then, the growth of dendritic textures was observed without a polarizer for all the PDI derivatives at around their clearing points (Figure 3a–c). At the same time, no optical texture appeared under crossed polarizers (Figure 3a–c). A similar behavior was seen for **H$_2$Pc** with slow cooling at 1.0 K/min (Figure 3d), while defect areas with homogeneous alignment were confirmed upon rapid cooling at 10 K/min (Figure S2). These microscopic observations indicate the strong homeotropic tendency for the hexagonally arranged discotic columns from all four compounds (Figure 3e). Interestingly, after the phase transition from a hexagonal to rectangular columnar mesophase at around 110 °C upon cooling, the micrograph of **PDI$_{C12/TEG}$** was almost unchanged, suggesting that the homeotropic orientation was kept upon the hexagonal–rectangular structural transformation.

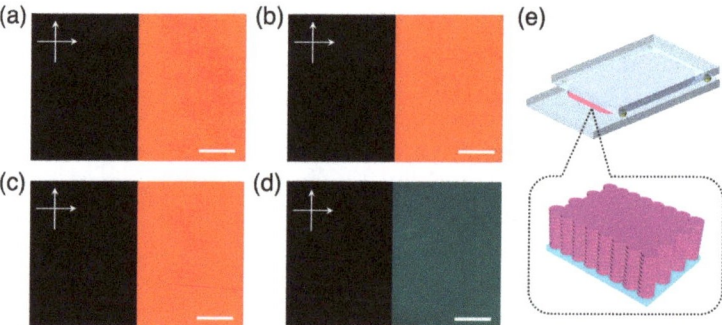

Figure 3. Crossed polarized (**left**) and optical (**right**) microscopy images of (**a**) **PDI$_{C12/C12}$**, (**b**) **PDI$_{C12/TEG}$**, (**c**) **PDI$_{TEG/TEG}$**, and (**d**) **H$_2$Pc** in glass sandwich cell without any treatment. (**a–d**) were taken at 222, 193, 168, and 181 °C, respectively, after cooling from their isotropic liquid phases at 1.0 K/min. Scale bars represent 200 μm. (**e**) Schematic illustration of LC samples in 5 μm thick sandwiched glass cell and homeotropic alignment of discotic columns formed in LC.

3.2. Orientation, Phase Transition Behavior, and Phase Structure of H_2Pc/PDI Mixtures

In order to confirm our hypothesis of the side-chain labeling strategy, 1:1 molar ratio mixtures of $H_2Pc/PDI_{C12/C12}$, $H_2Pc/PDI_{C12/TEG}$, and $H_2Pc/PDI_{TEG/TEG}$ were prepared and their phase behaviors were characterized. The three blend samples were loaded into a sandwich glass cell over 210 °C, and their optical textures were recorded upon cooling. Figure 4 shows optical micrographs with and without crossed polarizers and the dependence of the optical textures on the cooling rate. The mixture of $H_2Pc/PDI_{TEG/TEG}$ gave the most distinctive picture (Figure 4c,f). Independent of the cooling rate, the mixture clearly gave green and red color areas, which most likely correspond to the domains of H_2Pc and $PDI_{TEG/TEG}$, respectively. The hydrophobic H_2Pc and hydrophilic $PDI_{TEG/TEG}$ are immiscible with each other and segregated macroscopically [27]. In contrast, $H_2Pc/PDI_{C12/C12}$ and $H_2Pc/PDI_{C12/TEG}$ appear to have a homogeneous phase in the field of microscope view. Upon rapid cooling from their isotropic phases, fan-shaped textures appeared in POM upon Iso-to-LC phase transitions for both $H_2Pc/PDI_{C12/C12}$ and $H_2Pc/PDI_{C12/TEG}$ (Figure 4a,b). The presence of textures indicates a non-homeotropic alignment of columnar structures. In contrast, the cooling rate was set at 1.0 K/min, and the growth of dendritic textures was seen in optical microscopy without polarizers, but almost dark field images were obtained under crossed polarizers (Figure 4d,e). Although the dark area ratio in these blends was a bit smaller than their constituent compounds, the homeotropic alignment capability was confirmed by POM observations.

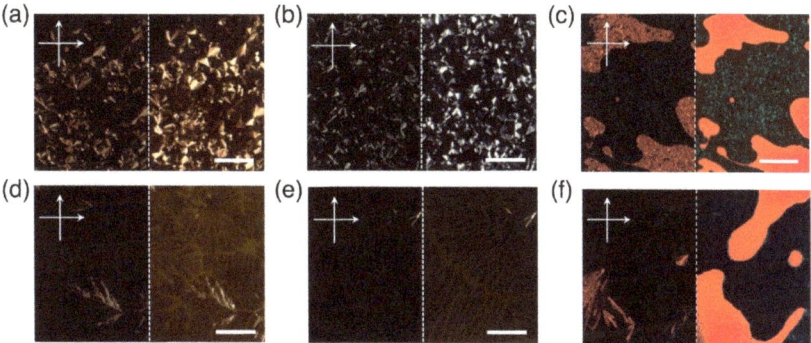

Figure 4. Crossed polarized (**left**) and optical (**right**) microscopy images of 1:1 molar ratio mixtures of (**a,d**) $H_2Pc/PDI_{C12/C12}$, (**b,e**) $H_2Pc/PDI_{C12/TEG}$, and (**c,f**) $H_2Pc/PDI_{TEG/TEG}$ in glass sandwich cell without any treatment. Images (**a–c**) were taken at 25 °C after rapid cooling from their isotropic melt. (**d–f**) were taken at 198, 200, and 161 °C, respectively, after cooling from their isotropic liquid phases at 1.0 K/min. Scale bars represent 200 μm.

The phase transition behaviors of the 1:1 molar ratio mixture of $H_2Pc/PDI_{C12/C12}$, $H_2Pc/PDI_{C12/TEG}$, and $H_2Pc/PDI_{TEG/TEG}$ were characterized by DSC. The DSC traces of $H_2Pc/PDI_{C12/C12}$ and $H_2Pc/PDI_{C12/TEG}$ implied phase transitions from a mesoscopically uniform material (Figure 5). In the blend of $H_2Pc/PDI_{C12/C12}$, the clearing point (202 °C on cooling) is between H_2Pc (180 °C) and $PDI_{C12/C12}$ (223 °C) (Figures 5a and S1), which is reasonable for molecularly miscible binary mixtures. In contrast, the clearing point of $H_2Pc/PDI_{C12/TEG}$ (205 °C on heating) is higher than those of H_2Pc (181 °C) and $PDI_{C12/TEG}$ (191 °C) (Figures 5b and S1). This pattern is quite rare and interesting to note—the LC phase of the blend is thermodynamically more stable than the parent columnar phases. We will discuss this phenomenon in more depth with the powder X-ray diffraction (PXRD) patterns (vide infra). In the blend of $H_2Pc/PDI_{TEG/TEG}$, the melting and clearing points of both the compounds are detected, though the clearing point at 192 °C is higher than that of H_2Pc (181 °C) (Figures 5 and S1). In other words, $H_2Pc/PDI_{TEG/TEG}$ affords the superimposed DSC chart of those of the constituent compounds. This is solely a sign of the macroscopic phase separation of **H₂Pc** and **PDI_TEG/TEG**, which is consistent with the POM images.

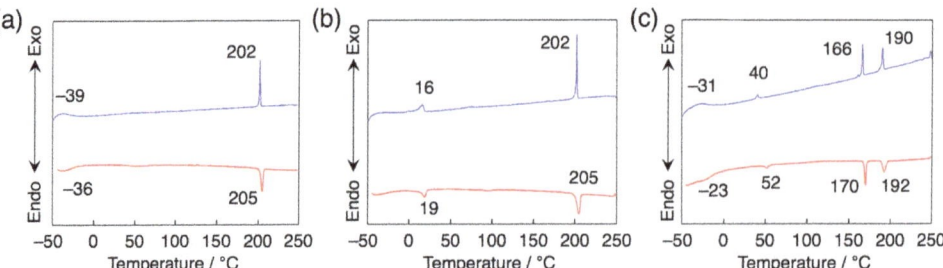

Figure 5. DSC traces of 1:1 molar ratio mixtures of (**a**) **H₂Pc/PDI_C12/C12**, (**b**) **H₂Pc/PDI_C12/TEG**, and (**c**) **H₂Pc/PDI_TEG/TEG** on 2nd heating/cooling cycle at 10 K/min.

Although the clearing points for **H₂Pc/PDI$_{C12/C12}$** and **H₂Pc/PDI$_{C12/TEG}$** are almost identical, the values of phase transition enthalpy inform that the LC phase structure and degree of miscibility are completely different between these mixtures. The LC-to-Iso phase transition enthalpy changes (ΔH) were evaluated from the second heating trace in DSC (Figure S1) and are 4.8, 17.3, and 8.1 kJ mol^{-1} for **H₂Pc**, **PDI$_{C12/C12}$**, and **PDI$_{C12/TEG}$**, respectively. The entropy changes upon these phase transitions (ΔS) can be estimated from the principle that Gibbs free energy is constant upon phase transition, i.e., $\Delta G = \Delta H - T\Delta S = 0$, where ΔG and T are Gibbs free energy change and absolute temperature. By substituting the evaluated ΔH and observed T into the above equation, the values of ΔS were estimated as 10.6, 34.8, and 17.4 J mol^{-1} K^{-1} for **H₂Pc**, **PDI$_{C12/C12}$**, and **PDI$_{C12/TEG}$**, respectively. These values well explain the relatively larger entropic gain of linear dodecyloxy chains upon phase transition from columnar mesophase to isotropic liquid. The values of ΔH and ΔS are calculated for the 1:1 molar mixtures of **H₂Pc/PDI$_{C12/C12}$** and **H₂Pc/PDI$_{C12/TEG}$** in a similar way, except that the average molecular weight of the two components is used for transforming the observed heat change into enthalpy values. The values of ΔS were estimated as 18.9 and 15.9 J mol^{-1} K^{-1} for **H₂Pc/PDI$_{C12/C12}$** and **H₂Pc/PDI$_{C12/TEG}$**, respectively. The value of 18.9 J mol^{-1} K^{-1} for **H₂Pc/PDI$_{C12/C12}$** is smaller than the averaged ΔS values calculated from those of the parent compound (22.7 J mol^{-1} K^{-1}), implying that the molecules are disordered in the observed columnar mesophase. For example, one column is composed of **H₂Pc** and **PDI$_{C12/C12}$** molecules. In contrast, the value of 15.9 J mol^{-1} K^{-1} for **H₂Pc/PDI$_{C12/TEG}$** is a bit larger than and even close to the averaged ΔS values of the parent compound (14.0 J mol^{-1} K^{-1}). This similarity in the entropy values indicates the possibility that **H₂Pc** and **PDI$_{C12/TEG}$** form their respective microdomains. The molecular motion in the microdomains upon the phase transition would be consistent with that in the bulk of the corresponding compounds, while that at the interfaces of the microdomains is relatively limited. In this case, the ΔS value is expected to be smaller than the average values speculated from those for the parent compounds. Considering that the phase transition temperatures for these blends and parent compounds are close and in the range of 181–224 °C, the above speculations would have a certain level of significance. As below, we will directly discuss the molecular packing structures in the mesophase for LC blends based on the PXRD measurements.

The molecular packing structures in the mesophases were studied by means of PXRD measurements. In the mesophase at 80 °C, the 1:1 molar ratio mixture of **H₂Pc/PDI$_{C12/C12}$** gave a diffraction pattern that is assignable to a hexagonal columnar phase with the lattice parameter of a = 32.1 Å (Figure 6a). Variable-temperature PXRD measurements elucidated that the hexagonal columnar mesophase was present at 30–200 °C (Figure S4). The parent hexagonal columnar mesophases of **H₂Pc** and **PDI$_{C12/C12}$** have a lattice parameter of a = ~32 Å and ~31 Å, respectively (Figures S3 and S7). The size matching of these two molecules may be one of the critical reasons for stabilizing a uniform hexagonal packing of mixed-stacked columns. With the clearing temperature information discussed in the DSC section, we conclude that the **H$_2$Pc/PDI$_{C12/C12}$** self-organized into molecularly miscible, entropically favored columns with hexagonal packing, as illustrated in Figure 7a. Interestingly, the mixture of **H$_2$Pc/PDI$_{C12/TEG}$** showed different behavior. Over 80 °C, the mixture formed a hexagonal columnar phase with a = ~32 Å (Figure S5). When being cooled down to 80 °C, the mixture changed its PXRD pattern to the superposition of those of **H$_2$Pc** and **PDI$_{C12/TEG}$** (Figure 6b), and similar superimposed patterns were also recorded at 50 and 30 °C (Figure S5). Namely, **H$_2$Pc** and **PDI$_{C12/TEG}$** are mesoscopically segregated but macroscopically miscible, as disclosed by PXRD and DSC measurements. The schematic illustration of **H$_2$Pc/PDI$_{C12/TEG}$** is shown in Figure 7b. Then, we tried to interpret the hexagonal columnar mesophase of **H$_2$Pc/PDI$_{C12/TEG}$** over 80 °C. Although a set of observed diffractions was assigned to a single hexagonal lattice, the (001) peak at d = ~3.4 Å, corresponding to the π-distance periodicity of **PDI$_{C12/TEG}$**, obviously appeared as similar to those at 30–80 °C. In addition, as mentioned in the DSC analysis earlier, the clearing temperature of the mixture at 205 °C is higher than those of the parent compounds. Having these results in mind, we consider that **H$_2$Pc** and **PDI$_{C12/TEG}$** mainly form self-sorted columns even over 80 °C but the average domain size may be decreased. The blend **H$_2$Pc/PDI$_{TEG/TEG}$** exhibited superimposed pattens of those of **H$_2$Pc** and **PDI$_{TEG/TEG}$** below 240 °C (Figures 6 and S6). These results are consistent with the macroscopic phase separation derived from the POM and DSC results. The illustration of macroscopically phase-separated columnar phases is shown in Figure 7c.

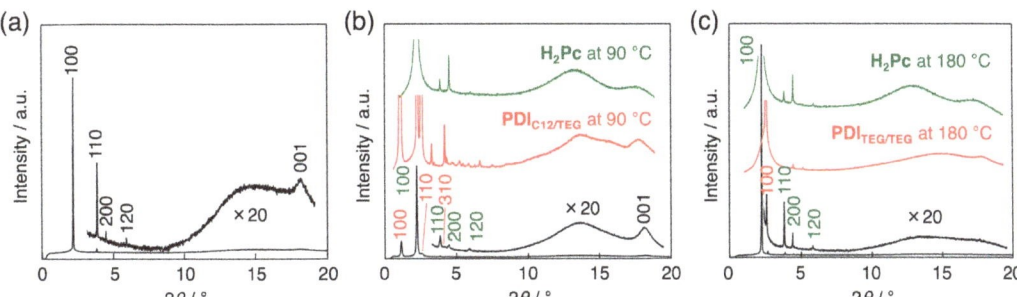

Figure 6. XRD patterns of 1:1 molar ratio mixtures of (**a**) **H$_2$Pc/PDI$_{C12/C12}$** at 80 °C, (**b**) **H$_2$Pc/PDI$_{C12/TEG}$** at 80 °C, and (**c**) **H$_2$Pc/PDI$_{TEG/TEG}$** at 160 °C. For comparison, the XRD patterns of the components for the blends are represented in (**b**,**c**).

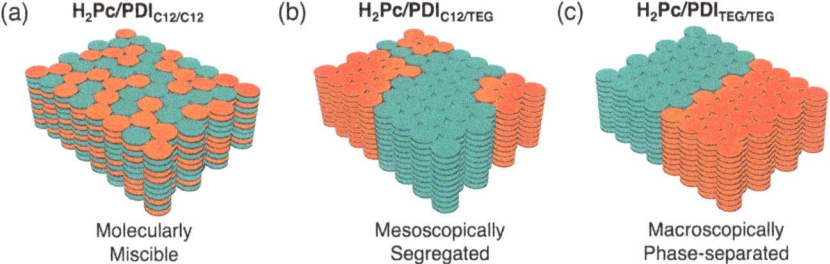

Figure 7. Schematic illustrations of proposed molecular assembly in columnar LC phases for (**a**) **H$_2$Pc/PDI$_{C12/C12}$**, (**b**) **H$_2$Pc/PDI$_{C12/TEG}$**, and (**c**) **H$_2$Pc/PDI$_{TEG/TEG}$**. Red and green disks represent corresponding H$_2$Pc and PDI molecules.

3.3. Intracolumnar Molecular Order in H_2Pc/PDI Mixtures

The intracolumnar molecular order in the mesophases at room temperature was investigated by absorption spectroscopy of the thin film of the 1:1 molecular blends. In diluted $CHCl_3$ solutions, both H_2Pc and $PDI_{C12/C12}$ are molecularly dispersed and show characteristic absorption at 600–750 nm and 400–550 nm, respectively, with strong vibronic coupling features (Figure 8a). In spin-coated LC films, these absorption bands become broad and blue-shifted due to the columnar assembly of molecules with π–π interactions (H-like aggregation). The spectra of $PDI_{C12/TEG}$ and $PDI_{TEG/TEG}$ in the films are essentially the same as that of $PDI_{C12/C12}$. Then, the spectra of the blend films were analyzed similarly. As expected, in the macroscopically phase-separated $H_2Pc/PDI_{TEG/TEG}$ blend film, the shape of the absorption spectra is almost the superposition of those of the parent LC films (Figures 8b and S8). The heterotropic interactions hardly work due to the limited area of the interfaces between H_2Pc and $PDI_{TEG/TEG}$. In the LC phase of $H_2Pc/PDI_{C12/C12}$, proposed as a molecularly miscible columnar phase, the absorption spectrum of the film is completely different from that of $H_2Pc/PDI_{TEG/TEG}$. The characteristic two intense absorption bands from H_2Pc and $PDI_{C12/C12}$ both show vibronic structures in the blend film, while these bands are broadened compared to their solution states (Figure 8b). This feature strongly indicates that homotropic molecular interactions in their columnar assembly are broken, supporting the proposed molecularly miscible columnar phase (Figure 7a). The film of the $H_2Pc/PDI_{C12/TEG}$ mixture afforded basically the superimposed spectrum of those of H_2Pc and $PDI_{C12/TEG}$. However, shoulder vibronic peaks at around 670–730 nm suggest that a small part of H_2Pc columnar assemblies is dissociated by the intercalation of $PDI_{C12/TEG}$. Thus, the picture of mesoscopically segregated self-sorted assembly as illustrated in Figure 7b may almost be correct, but the structural purity is less than perfect.

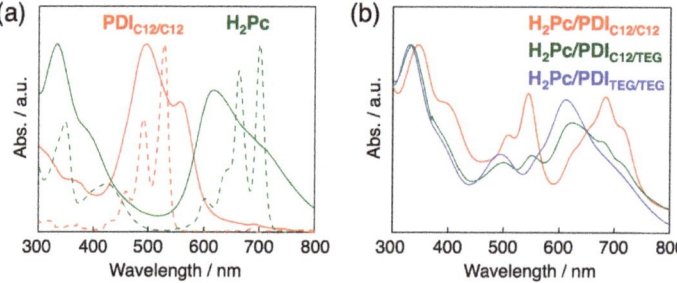

Figure 8. (a) Absorption spectra of H_2Pc (green) and $PDI_{C12/C12}$ (red) in spin-coated film (solid line) and in $CHCl_3$ (dotted line). (b) Absorption spectra of spin-coated film of $H_2Pc/PDI_{C12/C12}$ (red), $H_2Pc/PDI_{C12/TEG}$ (green), and $H_2Pc/PDI_{TEG/TEG}$ (blue).

4. Conclusions

Although nanosegregated, bicontinuous structures of electron-donating and accepting π-conjugated molecules have been recognized as important for photoconducting and photovoltaic properties, only the kinetic control of such nanostructures has been reported so far. We conceived the side-chain labeling strategy using hydrophobic/hydrophilic chains to induce the homotropic self-assembly of donor and acceptor molecules and demonstrated the preferential formation of donor/acceptor self-sorted columnar structures in thermodynamically stable LC binary mixtures. In this LC blend, the columnar mesophases of H_2Pc and PDI molecules are macroscopically miscible and uniform but mesoscopically segregated as evidenced by DSC and PXRD results. In addition, the intercalation of PDI (H_2Pc) to the H_2Pc (PDI) columns is minimally inhibited as supported by absorption spectroscopy. In a more comprehensive view, self-sorted nanostructures of binary mixtures are entropically unfavored in general, but the present work clarified that they can be accessed thermodynamically by self-assembly processes with the help of enthalpic interactions of immiscible side-chain pairs. Amphiphilic molecules—**PDI**$_{C12/TEG}$ in this work—induce mesoscopic phase separation and avoid macroscopic phase separation. This role is referred to as a compatibilizer in the research field of macromolecules [40]. While a small molecular compatibilizer has recently been reported [41], our work further extends the concept to the strategy of accessing self-sorted nanostructures. In future, the important subjects include the analysis and control of the size of donor and acceptor nano(micro)-domains, which will lead to the manipulation of photo and electronic functions originating from nanosegregated donor/acceptor blends.

Supplementary Materials: The following supporting information can be downloaded at: https://www.mdpi.com/article/10.3390/cryst13101473/s1, Figure S1: DSC traces of **H**$_2$**Pc**, **PDI**$_{C12/C12}$, **PDI**$_{C12/TEG}$, and **PDI**$_{TEG/TEG}$; Figure S2: Crossed polarized microscopy images of **H**$_2$**Pc** in glass sandwich cell; Figure S3: Variable-temperature XRD patterns of **H**$_2$**Pc**; Figure S4: Variable-temperature XRD patterns of 1:1 molar ratio mixture of **H**$_2$**Pc**/**PDI**$_{C12/C12}$; Figure S5: Variable-temperature XRD patterns of 1:1 molar ratio mixture of **H**$_2$**Pc**/**PDI**$_{C12/TEG}$; Figure S6: Variable-temperature XRD patterns of 1:1 molar ratio mixture of **H**$_2$**Pc**/**PDI**$_{TEG/TEG}$; Figure S7: Schematic illustrations of columnar hexagonal and rectangular phases with corresponding lattice parameters and primary diffractions.

Author Contributions: T.S. conceived and designed the experiments; T.S. performed the experiments; T.S. and K.K. analyzed the data; T.S. and M.S. wrote the manuscript draft; K.K. revised the manuscript draft. All authors have read and agreed to the published version of the manuscript.

Funding: This research was funded by JSPS KAKENHI numbers 20H02710 from the Japan Society for the Promotion of Science and a research grant from TEPCO Memorial Foundation.

Data Availability Statement: Not applicable.

Acknowledgments: The synchrotron radiation XRD experiments were performed at BL44B2 in SPring-8 with the approval of RIKEN. T.S. thanks the Leading Initiative for Excellent Young Researchers program by Ministry of Education, Culture, Sports, Science and Technology (MEXT), Japan.

Conflicts of Interest: The authors declare no conflict of interest.

References

1. Robeson, L.M. *Polymer Blends: A Comprehensive Review*; Hanser Publishers: Munich, Germany, 2007.
2. Paul, D.R. *Polymer Blends*; Elsevier: Amsterdam, The Netherlands, 2012; Volume 1.
3. Thomas, S.; Grohens, Y.; Jyotishkumar, P. *Characterization of Polymer Blends: Miscibility, Morphology and Interfaces*; John Wiley & Sons: Hoboken, NJ, USA, 2014.
4. Barlow, J.W.; Paul, D.R. Polymer blends and alloys—A review of selected considerations. *Polym. Sci. Eng.* **1981**, *21*, 985–996. [CrossRef]
5. Yu, G.; Gao, J.; Hummelen, J.C.; Wudl, F.; Heeger, A.J. Polymer photovoltaic cells: Enhanced efficiencies via a network of internal donor-acceptor heterojunctions. *Science* **1995**, *270*, 1789–1791. [CrossRef]
6. Dennler, G.; Scharber, M.C.; Brabec, C.J. Polymer-Fullerene Bulk-Heterojunction Solar Cells. *Adv. Mater.* **2009**, *21*, 1323–1338. [CrossRef]
7. Brabec, C.J.; Gowrisanker, S.; Halls, J.J.M.; Laird, D.; Jia, S.; Williams, S.P. Polymer-Fullerene Bulk-Heterojunction Solar Cells. *Adv. Mater.* **2010**, *22*, 3839–3856. [CrossRef]
8. Lee, C.; Lee, S.; Kim, G.-U.; Lee, W.; Kim, B.J. Recent Advances, Design Guidelines, and Prospects of All-Polymer Solar Cells. *Chem. Rev.* **2019**, *119*, 8028–8086. [CrossRef]
9. Xu, C.; Zhao, Z.; Yang, K.; Niu, L.; Ma, X.; Zhou, Z.; Zhang, X.; Zhang, F. Recent progress in all-small-molecule organic photovoltaics. *J. Mater. Chem. A* **2022**, *10*, 6291–6329. [CrossRef]
10. Gao, H.; Sun, Y.; Meng, L.; Han, C.; Wan, X.; Chen, Y. Recent Progress in All-Small-Molecule Organic Solar Cells. *Small* **2023**, *19*, e2205594. [CrossRef]
11. Wu, X.; Tam, T.L.D.; Chen, S.; Salim, T.; Zhao, X.; Zhou, Z.; Lin, M.; Xu, J.; Loo, Y.-L.; Leong, W.L. All-Polymer Bulk-Heterojunction Organic Electrochemical Transistors with Balanced Ionic and Electronic Transport. *Adv. Mater.* **2022**, *34*, e2206118. [CrossRef]
12. Cheng, S.-S.; Huang, P.-Y.; Ramesh, M.; Chang, H.-C.; Chen, L.-M.; Yeh, C.-M.; Fung, C.-L.; Wu, M.-C.; Liu, C.-C.; Kim, C.; et al. Solution-Processed Small-Molecule Bulk Heterojunction Ambipolar Transistors. *Adv. Funct. Mater.* **2014**, *24*, 2057–2063. [CrossRef]
13. Sugiyasu, K.; Kawano, S.; Fujita, N.; Shinkai, S. Self-Sorting Organogels with p-n Heterojunction Points. *Chem. Mater.* **2008**, *20*, 2863–2865. [CrossRef]

14. Molla, M.R.; Das, A.; Ghosh, S. Chiral induction by helical neigbour: Spectroscopic visualization of macroscopic-interaction among self-sorted donor and acceptor π-stacks. *Chem. Commun.* **2011**, *47*, 8934–8936. [CrossRef] [PubMed]
15. Prasanthkumar, S.; Ghosh, S.; Nair, V.C.; Saeki, A.; Seki, S.; Ajayaghosh, A. Organic Donor–Acceptor Assemblies form Coaxial p–n Heterojunctions with High Photoconductivity. *Angew. Chem. Int. Ed.* **2015**, *54*, 946–950. [CrossRef] [PubMed]
16. Draper, E.R.; Lee, J.R.; Wallace, M.; Jäckel, F.; Cowan, A.J.; Adams, D.J. Self-sorted photoconductive xerogels. *Chem. Sci.* **2016**, *7*, 6499–6505. [CrossRef] [PubMed]
17. Seki, A.; Yoshio, M.; Mori, Y.; Funahashi, M. Ferroelectric Liquid-Crystalline Binary Mixtures Based on Achiral and Chiral Trifluoromethylphenylterthiophenes. *ACS Appl. Mater. Interfaces* **2020**, *12*, 53029–53038. [CrossRef] [PubMed]
18. Zhang, C.; Nakano, K.; Nakamura, M.; Araoka, F.; Tajima, K.; Miyajima, D. Noncentrosymmetric Columnar Liquid Crystals with the Bulk Photovoltaic Effect for Organic Photodetectors. *J. Am. Chem. Soc.* **2020**, *142*, 3326–3330. [CrossRef] [PubMed]
19. Franca, L.G.; Dos Santos, P.L.; Pander, P.; Cabral, M.G.B.; Cristiano, R.; Cazati, T.; Monkman, A.P.; Bock, H.; Eccher, J. Delayed Fluorescence by Triplet-Triplet Annihilation from Columnar Liquid Crystal Films. *ACS Appl. Electron. Mater.* **2022**, *4*, 3486–3494. [CrossRef]
20. Yang, Z.; Li, J.; Chen, X.; Fan, Y.; Huang, J.; Yu, H.; Yang, S.; Chen, E.-Q. Precisely Controllable Artificial Muscle with Continuous Morphing based on "Breathing" of Supramolecular Columns. *Adv. Mater.* **2023**, *35*, e2211648. [CrossRef]
21. Nunes da Silva, F.; Marchi Luciano, M.; Stadtlober, C.H.; Farias, G.; Durola, F.; Eccher, J.; Bechtold, I.H.; Bock, H.; Gallardo, H.; Vieira, A.A. Columnar Liquid Crystalline Glasses by Combining Configurational Flexibility with Moderate Deviation from Planarity: Extended Triaryltriazines. *Chem. Eur. J.* **2023**, *29*, e202203604. [CrossRef]
22. Takahashi, H.; Kohri, M.; Kishikawa, K. Axially Polar-Ferroelectric Columnar Liquid Crystalline System That Maintains Polarization upon Switching to the Crystalline Phase: Implications for Maintaining Long-Term Polarization Information. *ACS Appl. Nano Mater.* **2023**, *6*, 10531–10538. [CrossRef]
23. Delage-Laurin, L.; Swager, T.M. Liquid Crystalline Magneto-Optically Active Peralkylated Azacoronene. *JACS Au* **2023**, *3*, 1965–1974. [CrossRef]
24. Zucchi, G.; Donnio, B.; Geerts, Y.H. Remarkable miscibility between disk- and lathlike mesogens. *Chem. Mater.* **2005**, *17*, 4273–4277. [CrossRef]
25. Zucchi, G.; Viville, P.; Donnio, B.; Vlad, A.; Melinte, S.; Mondeshki, M.; Graf, R.; Spiess, H.W.; Geerts, Y.H.; Lazzaroni, R. Miscibility between Differently Shaped Mesogens: Structural and Morphological Study of a Phthalocyanine-Perylene Binary System. *J. Phys. Chem. B* **2009**, *113*, 5448–5457. [CrossRef]
26. Sakurai, T.; Yoneda, S.; Sakaguchi, S.; Kato, K.; Takata, M.; Seki, S. Donor/Acceptor Segregated π-Stacking Arrays by Use of Shish-Kebab-Type Polymeric Backbones: Highly Conductive Discotic Blends of Phthalocyaninatopolysiloxanes and Perylenediimides. *Macromolecules* **2017**, *50*, 9265–9275. [CrossRef]
27. Thiebaut, O.; Bock, H.; Grelet, E. Face-on Oriented Bilayer of Two Discotic Columnar Liquid Crystals for Organic Donor-Acceptor Heterojunction. *J. Am. Chem. Soc.* **2010**, *132*, 6886–6887. [CrossRef] [PubMed]
28. Sakurai, T.; Shi, K.; Sato, H.; Tashiro, K.; Osuka, A.; Saeki, A.; Seki, S.; Tagawa, S.; Sasaki, S.; Masunaga, H.; et al. Prominent Electron Transport Property Observed for Triply Fused Metalloporphyrin Dimer: Directed Columnar Liquid Crystalline Assembly by Amphiphilic Molecular Design. *J. Am. Chem. Soc.* **2008**, *130*, 13812–13813. [CrossRef] [PubMed]
29. Lehmann, M.; Jahr, M.; Gutmann, J. Star-shaped oligobenzoates with a naphthalenechromophore as potential semiconducting liquid crystal materials? *J. Mater. Chem.* **2008**, *14*, 2995–3003. [CrossRef]
30. Sakurai, T.; Tsutsui, Y.; Choi, W.; Seki, S. Intrinsic Charge Carrier Mobilities at Insulator–Semiconductor Interfaces Probed by Microwave-based Techniques: Studies with Liquid Crystalline Organic Semiconductors. *Chem. Lett.* **2015**, *44*, 1401–1403. [CrossRef]
31. Sakurai, T.; Tsutsui, Y.; Kato, K.; Takata, M.; Seki, S. Preferential formation of columnar mesophases via peripheral modification of discotic π-systems with immiscible side chain pairs. *J. Mater. Chem. C* **2016**, *4*, 1490–1496. [CrossRef]
32. Funahashi, M. Anisotropic electrical conductivity of n-doped thin films of polymerizable liquid-crystalline perylene bisimide bearing a triethylene oxide chain and cyclotetrasiloxane rings. *Mater. Chem. Front.* **2017**, *1*, 1137–1146. [CrossRef]
33. Tant, J.; Geerts, Y.H.; Lehmann, M.; De Cupere, V.; Zucchi, G.; Laursen, B.W.; Bjornholm, T.; Lemaur, V.; Marcq, V.; Burquel, A.; et al. Liquid Crystalline Metal-Free Phthalocyanines Designed for Charge and Exciton Transport. *J. Phy. Chem. B* **2005**, *109*, 20315–20323. [CrossRef]
34. Kato, K.; Tanaka, Y.; Yamauchi, M.; Ohara, K.; Hatsui, T. A Statistical approach to correct X-ray response non-uniformity in microstrip detectors for high-accuracy and high-resolution total-scattering measurements. *J. Synchrotron Radiat.* **2019**, *26*, 762–773. [CrossRef]
35. Hatsusaka, K.; Ohta, K.; Yamamoto, I.; Shirai, H. Discotic liquid crystals of transition metal complexes, Part 30: Spontaneous uniform homeotropic alignment of octakis(dialkoxyphenoxy)phthalocyaninatocopper(II) complexes. *J. Mater. Chem.* **2001**, *11*, 423–433. [CrossRef]
36. Wang, J.; He, Z.; Zhang, Y.; Zhao, H.; Zhang, C.; Kong, X.; Mu, L.; Liang, C. The driving force for homeotropic alignment of a triphenylene derivative in a hexagonal columnar mesophase on single substrates. *Thin Solid Films* **2010**, *518*, 1973–1979. [CrossRef]
37. Osawa, T.; Kajitani, T.; Hashizume, D.; Ohsumi, H.; Sasaki, S.; Takata, M.; Koizumi, Y.; Saeki, A.; Seki, S.; Fukushima, T.; et al. Wide-Range 2D Lattice Correlation Unveiled for Columnarly Assembled Triphenylene Hexacarboxylic Esters. *Angew. Chem. Int. Ed.* **2012**, *51*, 7990–7993. [CrossRef] [PubMed]

38. Kobayashi, Y.; Muranaka, A.; Kato, K.; Saeki, A.; Tanaka, T.; Uchiyama, M.; Osuka, A.; Aida, T.; Sakurai, T. A structural parameter to link molecular geometry to macroscopic orientation in discotic liquid crystals: Study of metalloporphyrin tapes. *Chem. Commun.* **2021**, *57*, 1206–1209. [CrossRef] [PubMed]
39. Schweicher, G.; Gbabode, G.; Quist, F.; Debever, O.; Dumont, N.; Sergeyev, S.; Geerts, Y.H. Homeotropic and Planar Alignment of Discotic Liquid Crystals: The Role of the Columnar Mesophase. *Chem. Mater.* **2009**, *21*, 5867–5874. [CrossRef]
40. Self, J.L.; Zervoudakis, A.J.; Peng, X.; Lenart, W.R.; Macosko, C.W.; Ellison, C.J. Linear, Graft, and Beyond: Multiblock Copolymers as Next-Generation Compatibilizers. *JACS Au* **2022**, *2*, 310–321. [CrossRef]
41. Grimann, M.; Ueberschaer, R.; Tatarov, E.; Fuhrmann-Lieker, T. Phase Separation and Nanostructure Formation in Binary and Ternary Blends of Spiro-Linked Molecular Glasses. *J. Phys. Chem. B* **2020**, *124*, 5507–5516. [CrossRef]

Disclaimer/Publisher's Note: The statements, opinions and data contained in all publications are solely those of the individual author(s) and contributor(s) and not of MDPI and/or the editor(s). MDPI and/or the editor(s) disclaim responsibility for any injury to people or property resulting from any ideas, methods, instructions or products referred to in the content.

MDPI
St. Alban-Anlage 66
4052 Basel
Switzerland
www.mdpi.com

Crystals Editorial Office
E-mail: crystals@mdpi.com
www.mdpi.com/journal/crystals

Disclaimer/Publisher's Note: The statements, opinions and data contained in all publications are solely those of the individual author(s) and contributor(s) and not of MDPI and/or the editor(s). MDPI and/or the editor(s) disclaim responsibility for any injury to people or property resulting from any ideas, methods, instructions or products referred to in the content.

www.ingramcontent.com/pod-product-compliance
Lightning Source LLC
LaVergne TN
LVHW070559100526
838202LV00012B/517